Zoltan J Kiss

Quantum Engine

2011

Order this book online at www.trafford.com
or email orders@trafford.com

Most Trafford titles are also available at major online book retailers.

Printed in the United States of America.

ISBN: 978-1-4269-5782-6 (sc)
ISBN: 978-1-4269-5783-3 (e)

Library of Congress Control Number: 2011902483

Trafford rev. 03/25/2011

www.trafford.com

North America & international
toll-free: 1 888 232 4444 (USA & Canada)
phone: 250 383 6864 • fax: 812 355 4082

to all my friends, who helped me to keep going

Rotating discs experiments were carried out in Gabor Timar Workshop in Paks, Hungary, with Gabor Timar and Istvan Balogh. The stand of the experiment with devices and measurement tools were manufactured by the Timar Workshop. Publishing the book is a good occasion for expressing my thanks for their support and expertise.

Especial thanks to William (Keith) Hardwick for his permanent encouragement and support.

I am full of gratitude to my wife, my daughters and the whole family for helping me writing the books.

Executive Summary

Particle accelerator is producing electricity!

Gravitation, the sphere symmetrical expanding acceleration of *Earth* generates *blue shift*. Elementary processes – speeded up, close to *c* in particle accelerators – are communicating with the system of reference of the *Earth*. The *blue shift* impact of the *Earth* this way intensifies the neutron process of the element in acceleration. The intensified neutron process results in intensified proton process, intensified electron generation and electron surplus.

Particle accelerator with particles in motion with a speed close to the speed of light is a kind of *power plant*, generating electricity.

This is not an energy from nothing, rather transformation of the *blue shift* energy of the sphere symmetrical expanding acceleration of the *Earth* into electron flow.

The quantum energy impact of the *Earth* is available – a gift for us to use.

> This book is the continuation of the discussion on energy balance of relativity, the subject of the first book published in 2007 and about quantum energy and mass balance, the subject of the second book, published in 2009. This is why the book starts with section 24.
> A summary of the first two books is given in the first part of this book.
>
> *The Energy Balance of Relativity, the Theory of Event Concentration and Acceleration for Infinite Time*
>
> *Trafford Publishing 2007; ISBN 1-4251-1502-0*
> www.trafford.com/06-3261
>
> *Quantum Energy and Mass Balance, the Gift of the Earth*
>
> *Trafford Publishing 2009; ISBN 1-4251-9157-3*
> www.trafford.com/08-1547

***M**aterial world* is built up on permanent change, harmony and balance. The process is simple: transformation of mass into energy and re-transformation of energy into mass – sphere symmetrical expanding acceleration against sphere symmetrical accelerating collapse.

Proton is the sphere symmetrical expanding acceleration of mass from $\lim v = 0$, the quasi stationary status, up to speed $i = \lim a\Delta t = c$ – the transformation of mass into energy.
Neutron is the sphere symmetrical accelerating collapse of mass from *quantum entropy* status, speed of $i = \lim a\Delta t = c$ to $\lim v = 0$ – re-transformation of energy into mass.
The *electron* process is sphere symmetrical expanding acceleration of mass at constant $i = \lim a\Delta t = c$ speed for infinite time, approaching the status of *quantum entropy* – the *blue shift* drive of the transformation process.

The *entropy* principle ensures: the transformation and the re-transformation parts of the balance are never completely equal. The *entropy* principle also guaranties: *matter* is always in mass or energy status, but simultaneously never in both.

Mass-energy and energy-mass transformations happen in time. Change in time is: *intensity*. The *entropy* in the balanced cycles results in intensity difference – infinite small, the smallest possible – unresolved intensity imbalance or intensity reserve – *energy quantum*!

The accumulating intensity imbalance, the system of all energy quantum is the *Quantum System of Reference*.

The *Quantum System of Reference* is sensitive to mass or time change impact and reacts as a loaded *membrane*: Any impact is transferred in any direction with speed of *c* – the speed of light (the transmission of the impulse).

There are *no "flying" photons!* The *Quantum Membrane* is the one, which transfers the generated *quantum energy* impact.

Elements vary. They are – with their specific proton-neutron process intensity and electron process *blue shift* – the materialized appearances of the balanced mass-energy transformation. Any deviation from this mass-energy balance results in magnetic features, isotopes, alpha, beta, gamma, neutron radiation and other "particles".

There are infinite time systems existing in parallel. The intensity of events, measured as impact within systems of reference is function of the time flow. The same "absolute" event has different impact within the elementary world and within the system of reference of the *Earth*. (Our time system of reference on the *Earth* is infinite times slower than the time system of elements. It means, our measurements, relative to the elementary world are intensified in infinite times. Events within the elementary world with infinite long duration therefore might be measured as static particles within our *Earth* system of reference.)

Communication between systems of reference, their impact to each other depends on their intensity relations. Systems can communicate with each other if their time systems are of the same dimension.

Acceleration of elementary processes close to *c*, makes the communication with the system of reference of the *Earth* possible.

Earth gravitation generates *blue shift*. This *blue shift* impact can only be utilized within the elementary world, if elementary "particles" are speeded up. The closer their speed is to *c* the more intensive is the electron generation of the element in acceleration. Generation of electron surplus is result of the impact of *gravitation*.

Our future depends on energy. In order to escape from the energy trap and to find the way out, we have to appreciate the global rule: balanced change in time. It is important to make a note on this, because our energy generation practice today is mainly based on the destruction of the balance.

Quantum Engine is our future energy source!

Table of Content

I

SUMMARY
of
Book 1 and Book 2

Energy Balance of Relativity
and
Quantum Energy and Mass Balance

P.1

Part 1
Time relations of *systems of reference* in motion

P. 1.1

1.1
Motion with *constant speed*

In order to characterise the motion with $v=const$, we have to take a universal event and assess the impact of the motion on the description of the event. This universal event is a *light beam* across a system of reference in motion with $v=const$.

Two systems of reference are taken: *SORto* is supposed to be a stationary one, *SORtv* is supposed to be in motion with speed $v=const$ relative to *SORto* in direction, parallel with increasing axis yo. The speed of *SORtv* in direction, parallel to xo is *zero*. Both are understood as three-dimensional systems. For the simplicity of the projection we picture only two coordinates of the systems. They are respectively xo-yo and x-y.

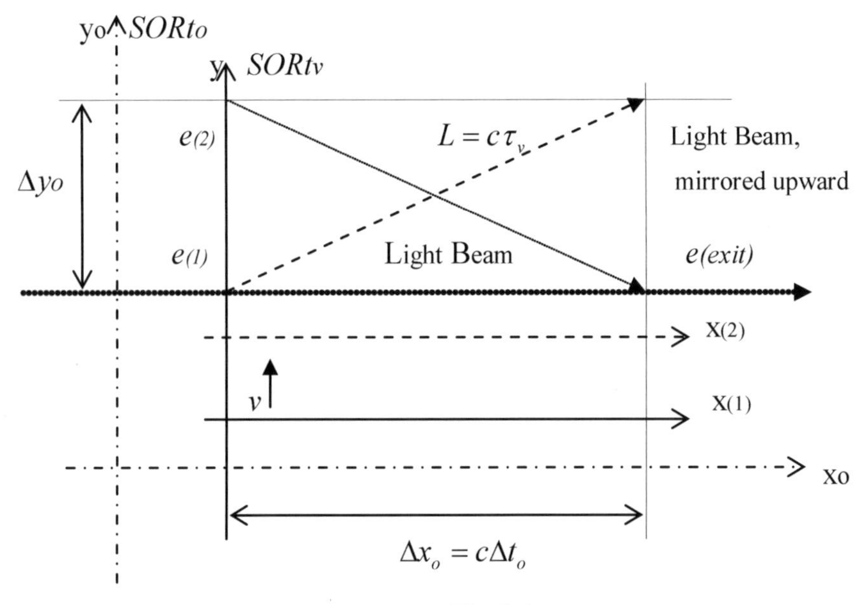

Fig. 1.1

Fig.1.1

The light signal enters *SORtv* at the spot, marked by *e(1)* at y, and crosses the system of reference not changing its direction. This direction is a straight line, parallel with axis xo, as observed within *SORto*. While the light beam crosses the systems, *SORtv* moves upward in direction yo with speed $v=const$ and makes a path of Δyo within *SORto*. As the result of this motion, axis $x(1)$ moves upward and takes position $x(2)$. This means that the spot of the access of the light signal *SORtv* moves together with the system upward into position *e(2)*.

We are looking for the description of this event within *SORtv*.
The light beam exits *SORtv* at distance of $\Delta xo = c\Delta to$.

The description of the light beam within *SORtv*, the direction, which would be observed within *SORtv* is a straight line *e(2) – e(exit)* declined under a certain angle to axis *x*, as it shown in Fig.1.1.

While the event is one and the same, its appearances in these two systems of reference are different:

in *SORto*, the observed trajectory is parallel with *xo*;

in *SORtv*, which is in motion, the observed trajectory is also a straight line, *but declined* on a certain angle to axis *x*.

(We may observe these trajectories, or we may not. We may be aware of the existence of any other system/s of reference, different to the one within which we may make our observations, or we may not. *Therefore*, it is better to use a word *description* for the characterisation of the event.)

We suppose that the light beam enters *SORtv* at time moment $t_o = 0$, measured in *SORto* and $\tau_v = 0$, measured in *SORtv*. The descriptions of the event in the two systems of reference are different.

The length of the path of the light beam in *SORto*, is: $\Delta x_o = c\Delta t_o$ or $dx_o = cdt_o$ 1A1

The description (or observation, if any) of the same event within *SORtv* is:

$$L = c\tau_v \quad \text{1A2}$$

τ_v is the time period while the light beam crosses *SORtv*: $\tau_v = \Delta\tau_v = \tau_v - 0$

The length of the path *SORtv* makes within *SORto* in direction *yo*, while the light beam crosses *SORtv* is the distance between spots *e(2)* and *e(1)* measured within *SORto*:

$$\Delta y_o = y_{o[e(2)]} - y_{o[e(1)]} \quad \text{1A3}$$

The distance, *SORtv* makes, measured within *SORtv* for $\Delta\tau_v = \tau_v - 0$, the duration of the event, also measured within *SORtv* is:

$$\Delta y = v\Delta\tau_v \quad \text{1A4}$$

Function for *SORtv* that satisfies these three (1A2, 1A3, 1A4) conditions is

$$L = c\tau_v = f(\Delta x_o; \Delta y); \quad \text{or} \quad L = f(dx_o; dy);$$

Speed is reciprocal, therefore: $\dfrac{\Delta y_o}{\Delta t_o} = v = \dfrac{\Delta y}{\Delta\tau_v}$; $\dfrac{dy_o}{dt_o} = v = \dfrac{dy}{d\tau_v}$;

$$dy_o = vdt_o; \quad dy = vd\tau_v; \quad \text{and} \quad dy = \frac{d\tau_v}{dt_o}dy_o$$

There are 2 variants for use:

1A5 variant (1): $\frac{d\tau_v}{dt_o} = \frac{1}{\sqrt{1-v^2/c^2}}$; and consequently $dy = \frac{dy_o}{\sqrt{1-(v^2/c^2)}}$;

1A6 variant (2): $\frac{dt_o}{d\tau_v} = \frac{1}{\sqrt{1-v^2/c^2}}$; and consequently $dy_o = \frac{dy}{\sqrt{1-(v^2/c^2)}}$;

We must come to the same formula as result of the deduced time relations for the systems of reference in relative motion with v=const!

The length of a curve in general, expressed by the coordinates in 1A4, is:

$$L = \int \sqrt{1+\left(\frac{dy}{dx_o}\right)^2}\, dx_o; \qquad ct_v = \int \sqrt{1+\left(\frac{dy}{dx_o}\right)^2}\, dx_o$$

Variant (1):

1A7

$$\frac{dy}{dx_o} = \frac{dy_o}{\sqrt{1-(v^2/c^2)}}\frac{1}{cdt_o} = \frac{dy_o}{dt_o}\frac{1}{c}\frac{1}{\sqrt{1-(v^2/c^2)}} = \frac{v}{c}\frac{1}{\sqrt{1-(v^2/c^2)}}$$

$$L = c\int \sqrt{1+\left(\frac{v}{c}\frac{1}{\sqrt{1-(v^2/c^2)}}\right)^2}\, dt_o; \qquad dL = cd\tau_v = \frac{c}{\sqrt{1-(v^2/c^2)}}dt_o$$

and results in

$$\boxed{\frac{d\tau_v}{dt_o} = \frac{1}{\sqrt{1-(v^2/c^2)}}}$$

[Variant (2): does not work!]

The motion speeds up the time flow. It flows faster within a system of reference in motion with v=const.

The question is: what is the result when the direction of the light beam (LB) is free?

The description of the event in *SORto* is: $LB = c\Delta t_o$

First we have to examine what kind of vector components relative to *xo* and *yo* the light beam may have, if α is the angle of the direction of the light signal relative to axis x and *xo*. Once we know the vector components relative to *xo* and *yo*, we are able to determine the impact of the motion of *SORtv* on the *description* of the event, the propagation of the light signal.

Can $LB \cdot \sin\alpha$ and $LB \cdot \cos\alpha$ be the vector components of the description of the light beam in *SORto* relative to axis *xo* and *yo* respectively? The answer is, obviously, *No*! Why?

Because

$LB \cdot \sin\alpha < \Delta x_o \;\; if \;\; \alpha \neq 90^\circ$ and $LB \cdot \cos\alpha < \Delta y_o \;\; if \;\; \alpha \neq 0^\circ$

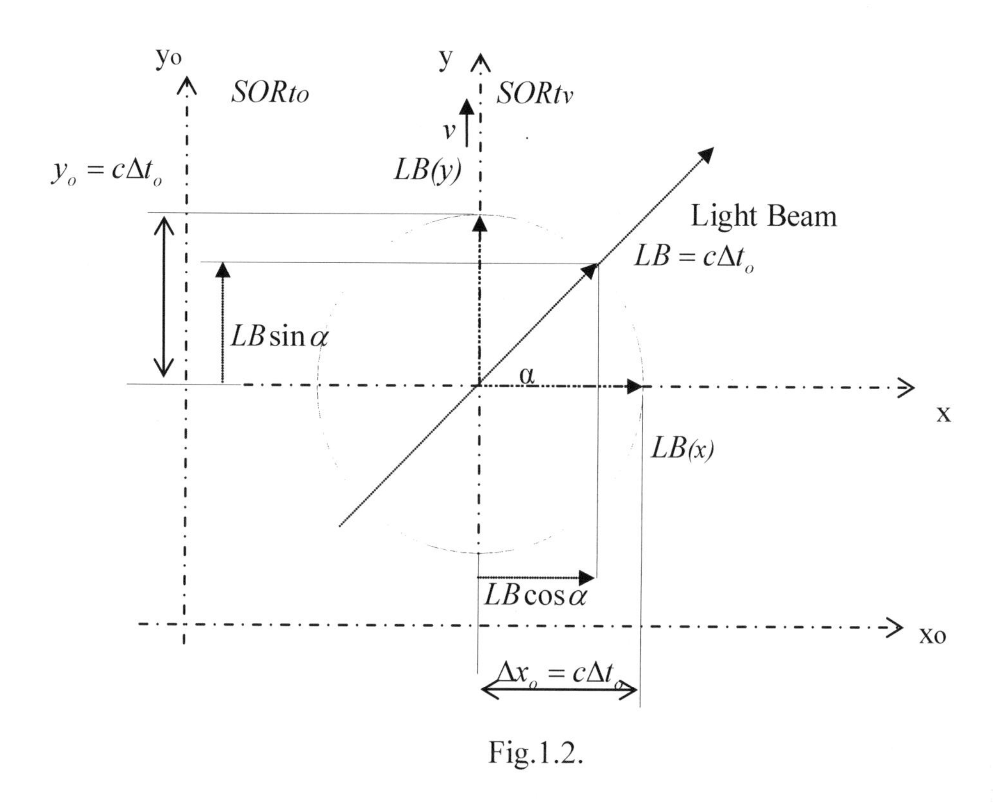

Fig.1.2.

Fig. 1.2

The vector components of the light beam must be equal to the light beam vector itself, otherwise either would be $c \neq const$, or $\Delta to \neq \Delta to$

$$LB = c\Delta t_o = \Delta x_o = \Delta y_o \tag{1A8}$$

The description of events in systems of reference in motion is independent of the direction of the light signal. It brings the fundamental consequence: the time relations are independent of the direction of the motion of the system of reference.

$$\tau_{(vparallel)} = \tau_{(vperpendicular)} = \tau_{(vanydirection)} = \tau_v = t_o \frac{1}{\sqrt{1-\left(v^2/c^2\right)}}; \text{ and}$$

$$\frac{d\tau_v}{dt_o} = \frac{1}{\sqrt{1-\left(v^2/c^2\right)}} \tag{1A9}$$

P. 1.2

1.2
Proof of time relations in Part 1.1, using *Einstein's* methodology
- with the measurement of space coordinates

Fig. 1.3

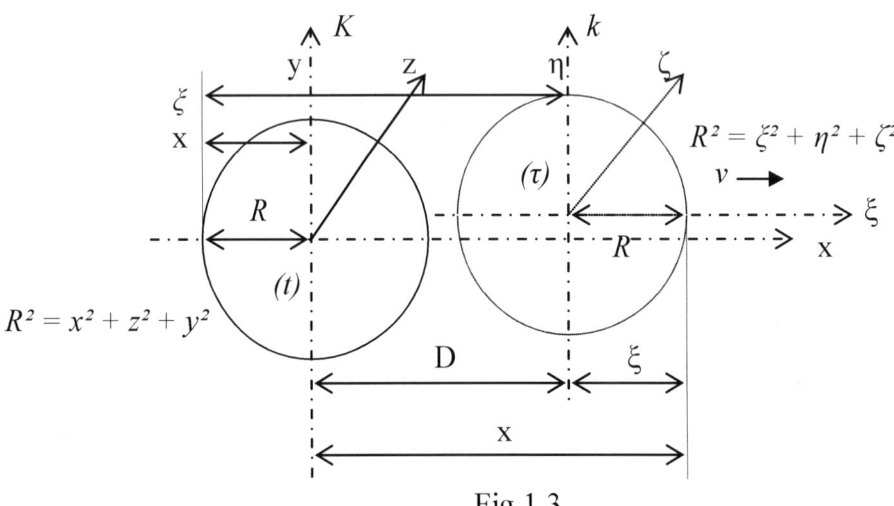

Fig.1.3

- System K is a stationary system, with coordinates *x, y* and *z;* the time measurement within the system is *t*;
- System *k*, with space coordinates ξ, η and ζ is in motion with velocity $v=const$ relative to system *K*, in direction, parallel to the increasing *x*;

This is not about evaluation *of an event* – but description of space coordinates.

In line with the transformation equations, we have to take into account the reciprocal character of the motion:

1B1

$$\frac{\tau_{(rel.m)}}{t} = \frac{1-(v/c)}{\sqrt{1-(v^2/c^2)}}; \qquad \frac{t_{(rel.m)}}{\tau} = \frac{1+(v/c)}{\sqrt{1-(v^2/c^2)}}$$

The meaning of both in fact is the same. Their mathematical descriptions however are clearly different. The use of "+"(positive) and the "–"(negative) signs depends on our free choice on the selection of the direction of the motion. We can write also write them as:

1B2

$$\frac{\tau_{(rel.m)}}{t} = \frac{1+(v/c)}{\sqrt{1-(v^2/c^2)}}; \qquad \frac{t_{(rel.m)}}{\tau} = \frac{1-(v/c)}{\sqrt{1-(v^2/c^2)}}$$

The equations are correct and their physical meaning is without change. We are going to bring them to a general meaning which excludes the significance of signs "+" and "–".

Summarising the left and right sides of 1B1 or 1B2: $\frac{\tau_{(rel.m)}}{t}+\frac{t_{(rel.m)}}{\tau}=\frac{2}{\sqrt{1-\left(v^2/c^2\right)}}$; 1B3

Summarising the $\frac{\tau_{(rel.m)}}{t}$ and the $\frac{t_{(rel.m)}}{\tau}$ parts of 1B1 and 1B2:

$$\frac{\tau_{(rel.m)}}{t}=\frac{1}{\sqrt{1-(v^2/c^2)}}\text{; and }\frac{t_{(rel.m)}}{\tau}=\frac{1}{\sqrt{1-(v^2/c^2)}} \quad \text{1B4}$$

The time relations for both systems involved are one and the same:

$$\frac{\tau_{(rel.m)}}{t}=\frac{t_{(rel.m)}}{\tau}=\frac{1}{\sqrt{1-\frac{v^2}{c^2}}} \quad \text{1B5}$$

For $\Delta t=t-0=t$
and $\Delta\tau=\tau-0=\tau$ time periods:

$$\frac{\Delta\tau_{(rel.m)}}{\Delta t}=\frac{\Delta t_{(rel.m)}}{\Delta\tau}=\frac{1}{\sqrt{1-\frac{v^2}{c^2}}} \quad \text{1B6}$$

1.3.
Proof of time relations in Part 1.1, using Einstein's methodology
- with the use of systems of reference taken in rotation

P.
1.3

In a space which is free of gravitational fields, we are taking two systems of reference in Fig.1.4, as in *Einstein's* classical example of his paper on "The foundation of the general theory of relativity" of 1916: a Galilean system of reference *Ko* and also *K*, a system of reference in uniform rotation relative to *Ko*. The origin *'o'* of both systems and their axes of *zo* and *z* permanently coincide.

We will show that, while space coordinates and time measurement cannot indeed be *projected* in conventional way, the main concern is not about the tools of projection rather the approach used.

Independently of whether a stationary system like *Ko* does exist at all, or does not, the rotation is *not uniform*. The values of the angular speed of the rotation of *K*, measured within *Ko* and measured within *K* itself are different.

Why?

Because *v*, the relative speed between the coordinates or positions of the two systems of reference in relative motion must be reciprocal. But *R*, the measured distance of a radius of and in the system of reference in motion at peripheral speed *v* of the rotating *K* differs from that of *Ro*, of the same rotation of *K*, but measured within *Ko,* the stationary system of reference.

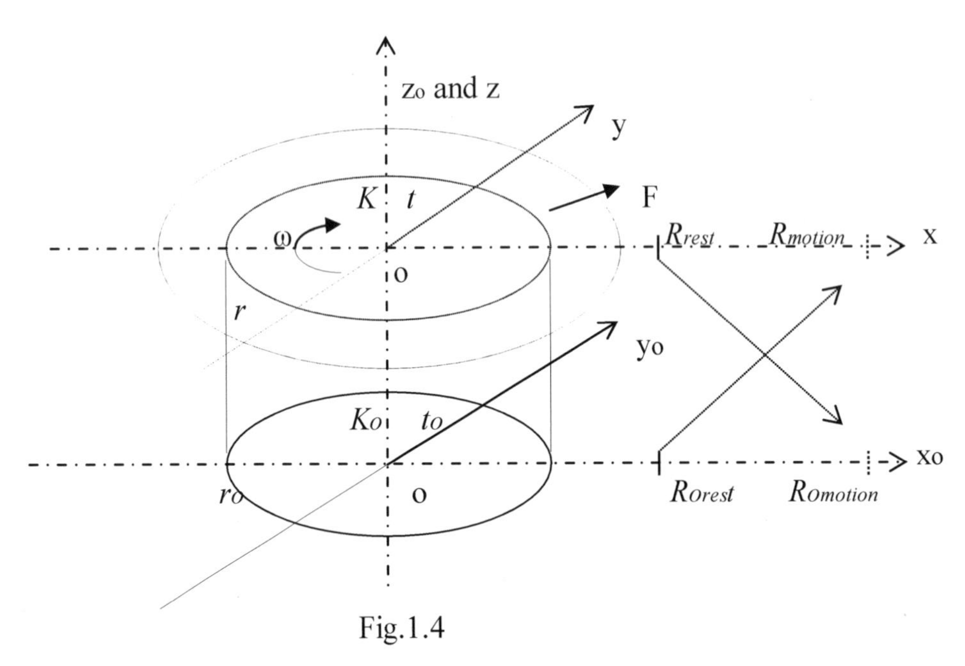

Fig. 1.4

Fig.1.4

The angular speed of the position or the coordinate with peripheral speed v, of the rotation of K measured in K is

$$\omega_R = \frac{v}{R} \text{ while the angular speed, measured in } Ko \text{ is } \omega_o = \frac{v}{R_o} \text{ and } \omega_R \neq \omega_o$$

The reciprocal character of the motion shall be taken into account. Without energy considerations of any kind, there is no reason to consider *a priori* one of the systems as *stationary* and the other as in motion, while both are equal parts of the same relation.

From the transformation equation: $c^2\Delta t_o^2 = c^2\Delta t_R^2 - v^2\Delta t_R^2$

1C1

$$v = \omega_R R \quad \text{and} \quad c^2\Delta t_o^2 = c^2\Delta t_R^2 - \omega_R^2 R^2 \Delta t_R^2$$

The angular speed at this circumference of radius R is:

1C2

$$\omega_R = \frac{2\Pi}{\Delta t_R} \quad \text{and} \quad R = \frac{c}{2\Pi}\sqrt{\Delta t_R^2 - \Delta t_o^2}$$

(R is measured within K, the system of reference in rotation at position or coordinate of mass point with peripheral speed v; ω_R is the angular speed of the rotation at this radius, measured within the system of reference in rotation; Δto is the period of time measured in Ko, the stationary system of reference, during the system of reference at a circumference with speed v of K makes a full spin.)

The meaning of the formula is: This is the position of the system of reference of a mass point in motion, which completes a full cycle in rotation for time period Δt. It is measured in the system of reference in motion at *distance* R from the centre of rotation.

$\Delta t_o = \frac{1}{c}\sqrt{c^2\Delta t_R^2 - D^2\Pi^2}$ and since $\Delta tR = \frac{2\Pi}{\omega}R$ it is equal to

$$\Delta t_o = \frac{2\Pi}{c}\sqrt{\frac{c^2}{\omega_R} - R^2} \quad \text{1C3}$$

Since the peripheral speed at radius R of the rotating system of reference is $v = \omega_R R$

$$\Delta t_o = \frac{2\Pi}{c}R\sqrt{\frac{c^2}{v^2} - 1} \quad \text{which is equal to} \quad \boxed{\Delta t_o = \Delta t_R\sqrt{1 - \frac{v^2}{c^2}}} \quad \text{1C4}$$

it gives the time relation of systems of reference in relative motion

Since: $v = \omega_R R$; $\omega_R = \frac{2\Pi}{\Delta t_R}$; $R = \Delta t_R \frac{v}{2\Pi}$; $\Delta t_o = \frac{2\Pi}{c}\frac{v}{2\Pi}\Delta t_R \frac{\sqrt{c^2 - v^2}}{v}$

The meaning of 1C4 is fundamental:

- The motion speeds up the time flow. *It proves, contrary to Einstein's time formula that the time flows faster in systems of reference in motion!* The strength of this deduction is that only the transformation equations and the tools of Euclidian geometry were used.
- It gives the meaning of time: *no motion (no event) = time can not be defined*:

$$\omega_R = \frac{v}{R};\quad \Delta t_o = \frac{2\Pi}{c}R\frac{\sqrt{c^2 - v^2}}{v};\ \text{and}\quad \boxed{\Delta t_o = \frac{2\Pi}{c}\cdot\frac{1}{\omega_R}\sqrt{c^2 - (w_R R)^2}} \quad \text{1C5}$$

It allows $R = 0$, but $\omega R \neq 0$! If $\omega R = 0$ it has no meaning. Consequently, time parameters can only be defined if motion (event) is present.

With reference to the time relation: $\Delta t_R = \frac{\Delta t_o}{\sqrt{1 - (v^2/c^2)}}$, but $\Delta t_o = \frac{2\Pi}{c}R\sqrt{\frac{c^2}{v^2} - 1}$ and 1C6

the radius is:
measured in the system of reference in motion

$$\boxed{R = \frac{v\Delta t_R}{2\Pi} = \frac{v\Delta t_o}{2\Pi}\cdot\frac{1}{\sqrt{1 - \frac{v^2}{c^2}}}} \quad \text{1C7}$$

Meaning: *motion speed up the time flow and expands the space!*

P. 1.4

1.4
Resolution of *Einstein's* time paradox

The *time paradox* is the reciprocal time relation of two systems of reference, relative motion to each other with constant speed. There is no way to decide which system is in motion relative to the other, the time system of which is in time delay relative to the other.
The description of the relation of the two systems shall take this fact into account.

Lorentz' transformation, the classical description of the event is:

1D1
$$\Delta\tau_{motion} = \frac{\Delta t - (vx/c^2)}{\sqrt{1-(v^2/c^2)}}; \quad \text{and} \quad \xi_{motion} = \frac{x - v\Delta t}{\sqrt{1-(v^2/c^2)}};$$

The original relation of the two systems:

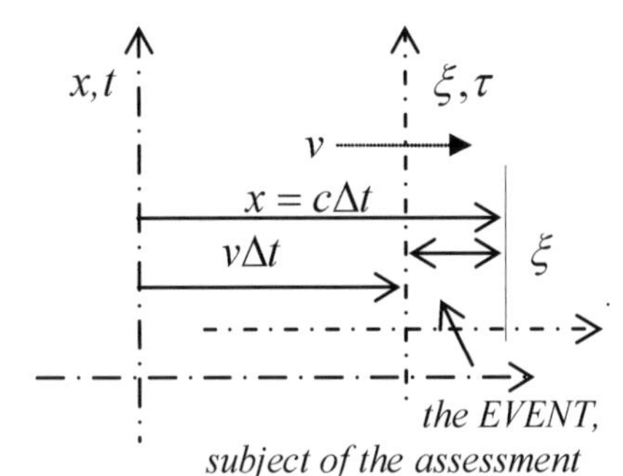

happens while ξ,τ makes distance ξ

Einstein's interpretation however represents the following case:

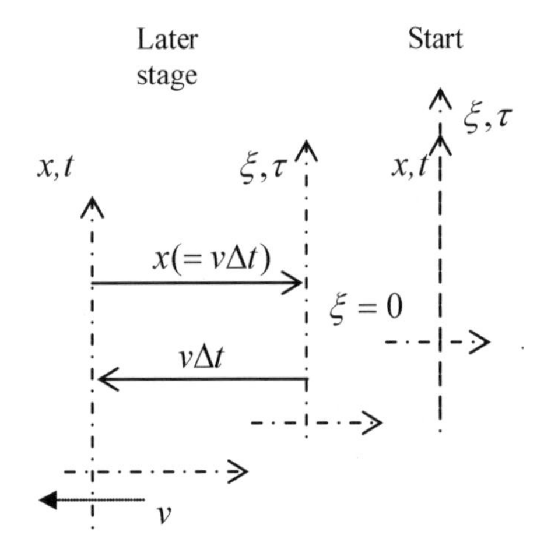

System of reference (SoR) of ξ,τ is in motion within x,t with speed v.

Einstein had made a note about the reciprocal character in his assessment, but there is no reference to this fact in his formulas.
Instead He substitutes in the formula

$x = v\Delta t$ *instead of* $x = c\Delta t$,
as measured event or subject.

This is rather strange, because $x = c\Delta t$ would be the end of the measured event or coordinate within the system of reference of x,t. The case, with this particular substitution becomes like this:

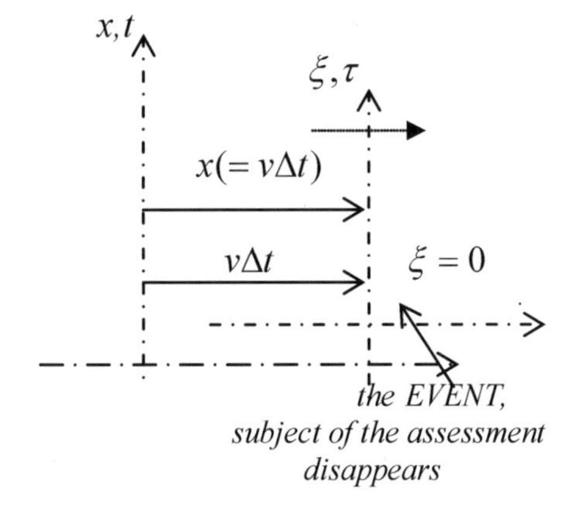

continuation of the columns on the previous page

SoR of x,t is moving away from SoR of ξ,τ with $v=c$

$x=v\Delta t$ the *EVENT* or space coordinate

gives $\xi=\dfrac{x-x}{\sqrt{1-(v^2/c^2)}}=0$

The motion is reciprocal, but $\xi=0$ means in this case that SoR of ξ,τ is taken as stationary!

And the solutions gives the result as:

$$\Delta\tau=\frac{\Delta t-(vx/c^2)}{\sqrt{1-(v^2/c^2)}}=$$

$$\frac{\Delta t-\Delta t(v^2/c^2)}{\sqrt{1-(v^2/c^2)}}=\Delta t\sqrt{1-\frac{v^2}{c^2}} \qquad \text{1D2}$$

which is convergent with all relevant equations, the Doppler formula and the Minkovsky's space-time interval indeed, BUT there is NO EVENT in System of Reference of ξ,τ to count with.

Einstein's solution represents the case on the left hand side:

event x within the System of Reference of x,t in motion with $v=c$.

If SoR of ξ,τ taken as stationary,

$$\Delta\tau_{rest}=\frac{\Delta t-(vx/c^2)}{\sqrt{1-(v^2/c^2)}}=\frac{\Delta t-\Delta t(v^2/c^2)}{\sqrt{1-(v^2/c^2)}}=\Delta t_{motion}\sqrt{1-\frac{v^2}{c^2}}\text{; and}$$

$$\Delta t_{motion}=\frac{\Delta\tau_{rest}}{\sqrt{1-(v^2/c^2)}} \qquad \text{1D3}$$

which has in this case only theoretical meaning since $\Delta t_{motion}=\Delta\tau_{rest}=0$ (as should be because $v=c$)

The correct solution is:

The reciprocal approach gives the reciprocal description of the *Lorentz's* transformation equations:

$$\tau_{motion}=\frac{t-(vx/c^2)}{\sqrt{1-(v^2/c^2)}}\text{; and }\xi_{motion}=\frac{x-vt}{\sqrt{1-(v^2/c^2)}}\text{; and also} \qquad \text{1D4}$$

$$t_{motion}=\frac{\tau-(v\xi/c^2)}{\sqrt{1-(v^2/c^2)}}\text{; and }x_{motion}=\frac{\xi-v\tau}{\sqrt{1-(v^2/c^2)}} \qquad \text{1D5}$$

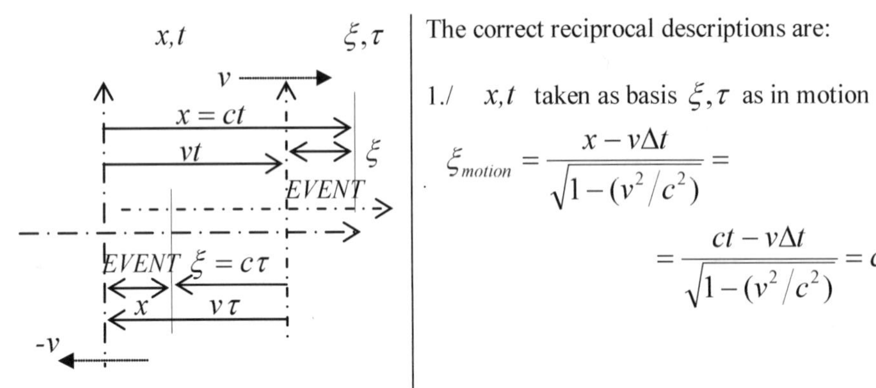

The correct reciprocal descriptions are:

1./ x,t taken as basis ξ,τ as in motion

$$\xi_{motion} = \frac{x - v\Delta t}{\sqrt{1-(v^2/c^2)}} = \frac{ct - v\Delta t}{\sqrt{1-(v^2/c^2)}} = c\Delta\tau_{motion}$$

2./ ξ,τ taken as basis, x,t as in motion

$$x_{motion} = \frac{\xi - (-v)\Delta\tau}{\sqrt{1-(v^2/c^2)}} = \frac{\xi + v\Delta\tau}{\sqrt{1-(v^2/c^2)}} = \frac{c\Delta\tau + v\Delta\tau}{\sqrt{1-(v^2/c^2)}} = c\Delta t_{motion}$$

The selection of (+) positive direction of the motion is our free choice and the equations also may be written as:

$$\xi_{motion} = \frac{ct + v\Delta t}{\sqrt{1-(v^2/c^2)}} = c\Delta\tau_{motion} \quad \text{and} \quad x_{motion} = \frac{c\Delta\tau - v\Delta\tau}{\sqrt{1-(v^2/c^2)}} = c\Delta t_{motion}$$

Summarising the equations: $2\frac{\Delta\tau_{motion}}{\Delta t} = \frac{2}{\sqrt{1-(v^2/c^2)}}$; and $2\frac{\Delta t_{motion}}{\Delta\tau} = \frac{2}{\sqrt{1-(v^2/c^2)}}$

as well as $\frac{\Delta\tau_{motion}}{\Delta t} + \frac{\Delta t_{motion}}{\Delta\tau} = \frac{1-(v/c)+1+(v/c)}{\sqrt{1-(v^2/c^2)}} = \frac{2}{\sqrt{1-(v^2/c^2)}}$

1D6 The time relations are <u>equal</u> and <u>reciprocal</u>: $\frac{\Delta\tau_{motion}}{\Delta t} = \frac{\Delta t_{motion}}{\Delta\tau} = \frac{1}{\sqrt{1-(v^2/c^2)}}$

The motion itself and its direction have physical impact on the frequency relations. The energy component of the motion is important.
Time relations are *no more reciprocal.*

1D7 $1\mp\frac{v}{c}$ components, introduced above give <u>*natural*</u> *correction*: $f_v = f_o\sqrt{\frac{1\pm\frac{v}{c}}{1\mp\frac{v}{c}}}$

There are two sides of the same energy transfer:

- By granting energy to the radiation, the energy of the system of reference in motion must also be corrected. The correction results in less kinetic energy, in slower speed of the system of reference of the collision. With slower speed the consequence of the change of the time relations is the increase of the frequency. This change is reflected by the $[1-(v/c)]$ correction.
- Due to the energy transfer from the radiation, the kinetic energy of the system of reference in motion increases. With increasing speed the time relations will change and, as a consequence, the frequency will decrease. The balance is made through $[1+(v/c)]$.

The results are in full compliance with the *Doppler* formula.

Alongside with the experienced frequency relations, consequence of the impact of external energy, the *sphere symmetrical expanding acceleration* and *accelerating collapse*, processes of the internal mass-energy balance can be fully described by the comprehensive *Doppler* formula.

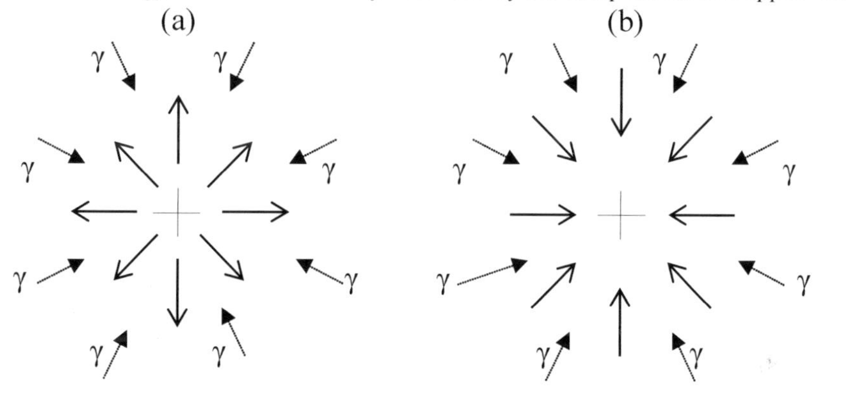

Fig.1.5

Fig. 1.5

a./ Collision from in front of the motion:

$$\frac{f_\alpha}{f} = \sqrt{\frac{1-(v/c)}{1+(v/c)}}; \quad \Delta E = (f - f_\alpha)H = \Delta f H$$

speed difference between the *Quantum Membrane* and the mass *system of reference is*:

$$v = c - \upsilon$$

$\Delta E = 0$ *no energy transfer*	$\Delta f = 0$	$\frac{df_\alpha}{df} = 1$	$\sqrt{\frac{1-(v/c)}{1+(v/c)}} = 1$	$v = 0;\ \upsilon = c$ $f_\alpha = f$
$\Delta E > 0$ *radiation gives off energy*	$\Delta f > 0$	$\frac{df_\alpha}{df} < 1$	$f_\alpha = f\sqrt{\frac{1-(v/c)}{1+(v/c)}}$	$0 < \upsilon < i = \lim c$ $f_\alpha < f$
$\Delta E < 0$ *mass system of reference gives off energy*	$\Delta f < 0$	$\frac{df_\alpha}{df} > 1$	$f_\alpha = f\sqrt{\frac{1+(v/c)}{1-(v/c)}}$	$\upsilon = i = \lim c$ $f_\alpha > f$

b./ Collision from behind of the motion: $\frac{f_\omega}{f} = \sqrt{\frac{1+(v/c)}{1-(v/c)}}$; $\Delta E = (f - f_\omega)H = \Delta f H$

speed difference between the Quantum Membrane and the mass *system of reference*:

$v = c - \upsilon$

$\Delta E = 0$ *no energy transfer*	$\Delta f = 0$	$\frac{df_\omega}{df} = 1$	$\sqrt{\frac{1+(v/c)}{1-(v/c)}} = 1$	$v = c;\ \upsilon = 0$ $f_\omega = f$
$\Delta E < 0$ *mass system of reference gives off energy*	$\Delta f < 0$	$\frac{df_\omega}{df} > 1$	$f_\omega = f\sqrt{\frac{1+(v/c)}{1-(v/c)}}$	$i = \lim c > \upsilon > 0$ $f_\omega > f$
$\Delta E > 0$ *radiation gives off energy*	$\Delta f > 0$	$\frac{df_\omega}{df} < 1$	$f_\omega = f\sqrt{\frac{1-(v/c)}{1+(v/c)}}$	$\upsilon = i = \lim c$ $f_\omega < f$

P. 1.5

1.5
Acceleration

SORto and SOR*ta* are understood as three-dimensional inert systems. For the simplicity of the projection we picture only two coordinates of the systems. They are respectively *xo-yo* and *x-y*. We suppose that *SORta* is accelerating within SOR*to*. SOR*to* is supposed to be a stationary system. The two systems of reference are shown in Fig.1.6.

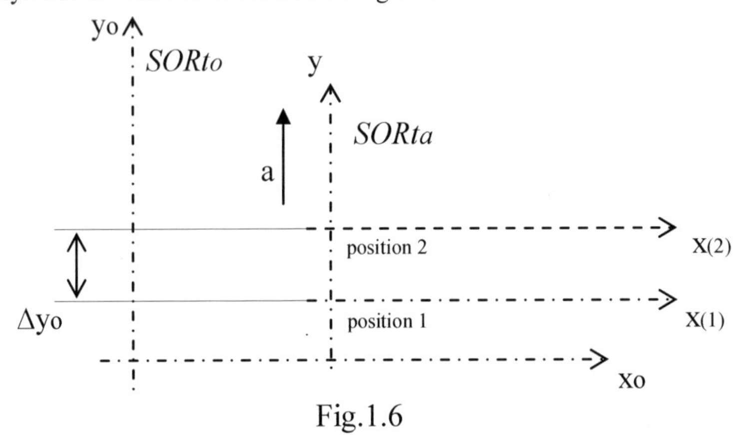

Fig. 1.6

Fig.1.6

SORta is accelerating with *a=const* in direction *yo*. For a certain period of time it moves from *position 1* to *position 2* within *SORto*. Axis *y* is parallel with *yo*. Axis *xo* and *x* are also parallel with each other.

We are looking for the *path* of the acceleration of *SORta* within *SORto*.

We divide the process of the acceleration into an infinite number of equal and *as small as possible* time periods.

Δt_o has been taken as infinitely small and the smallest possible. We suppose that the values of the speed will remain unchanged within the small as possible time periods and will increase only in certain time points at the beginning of each period. The values of the speed in these actual time points are: $v_{(n-1)}$, $v_{(n)}$ and $v_{(n+1)}$ accordingly. At time point $t_{o(0)}=0$ the value of the speed is $v_{(0)}=0$.

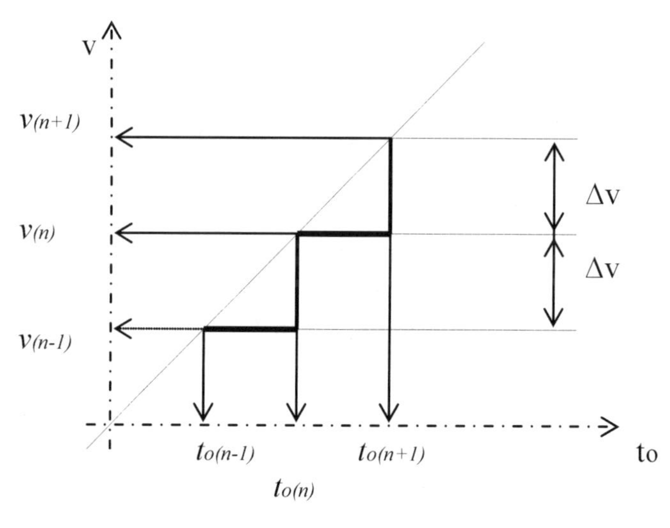

Fig.1.7

Fig 1.7

The accelerating process has been divided into an infinite number of equal periods, with, for each period, constant speed. The speed grows in time points $t_{o(n-1)}$, $t_{o(n)}$ and $t_{o(n+1)}$. It is constant in each period, but period by period of higher value, as follows:

$t_{o(n+1)} - t_{o(n)} = \Delta t_o$ and for this period the speed is $v_{(n)} = const$

$t_{o(n)} - t_{o(n-1)} = \Delta t_o$ and for this period the speed is $v_{(n-1)} = const$ and so on

and $v_{(n-1)} < v_{(n)} < v_{(n+1)}$

With reference to Section 1.1, $\Delta\tau$ the time period in *SORta* relates to Δt_o, the smallest possible time period, measured in *SORto*, as:

$$d\tau = dt_o \frac{1}{\sqrt{1 - v^2/c^2}} \quad \text{1E1}$$

The distance, *SORta* makes for the smallest possible time period within *SORto*, measured both the time and distance in *SORto* is:

$$\frac{dy_o}{dt_o} = v \quad \text{and} \quad v = at_o \quad dy_o = at_o dt_o \quad \text{1E2}$$

The speed is reciprocal $\dfrac{dy}{d\tau} = v$ therefore:

1E3 $$dy = vdt_o \frac{1}{\sqrt{1-(v^2/c^2)}}; \quad \text{and} \quad dy = at_o \frac{1}{\sqrt{1-(a^2t_o^2/c^2)}} dt_o$$

The length of the path of the accelerating motion of *SORta* in direction *yo* in Fig.1.6 (from position 1 to 2) *measured* in *SORta* is

1E4 $$\Delta y = s_a = \int at_o \frac{1}{\sqrt{1-(a^2t_o^2/c^2)}} dt_o$$

1E5 $$s_a = \frac{c^2}{a}\left(1-\sqrt{1-\frac{a^2t_o^2}{c^2}}\right) \quad \text{or} \quad s_a = \frac{c^2}{a} - \frac{c^2}{a}\sqrt{1-\frac{a^2t_o^2}{c^2}}$$

The relation of the paths of the accelerating motion of *SORta, at a certain time moment*, expressed or measured in *SORta*:

$$dy = \frac{dy_o}{\sqrt{1-(a^2t_o^2/c^2)}}$$

The distance in *SORto* is $dy_o = at_o dt_o$ gives the classical formula

$$\Delta y_o = s_o = \frac{1}{2} at_o^2$$

Acceleration differs from the motion with *v=const*, where the relation of the systems in motion to each other is relative and equal. Through its force effect we are always in a position to identify which system is in acceleration. Light beam crosses both systems in a direction perpendicular to the direction of the acceleration.

Fig 1.8

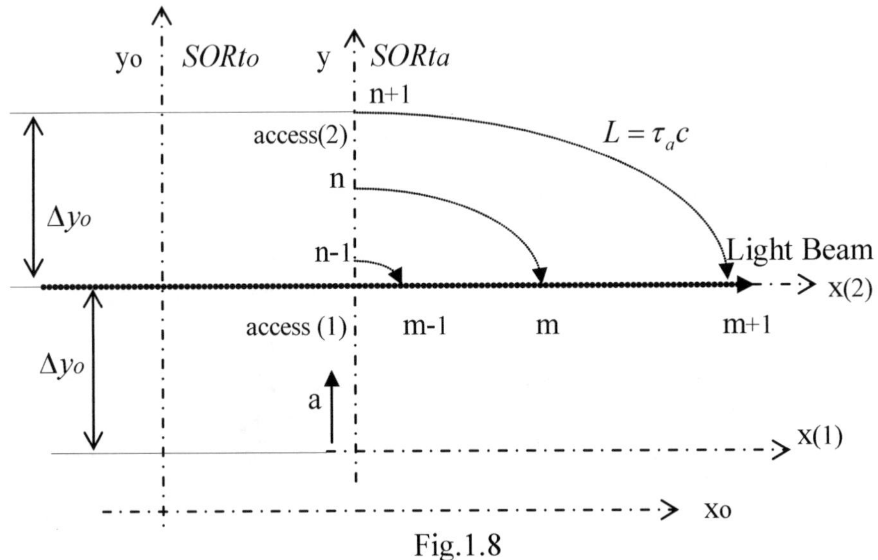

Fig.1.8

SORta makes a distance of Δyo in *SORto* while the light beam crosses *SORta*. The access spot of the light signal into *SORta* moves obviously the same distance in the direction of the acceleration.

During this period, the supposed traveller inside the system would observe different curves in different time periods that are shown in Fig.1.8 as curves *(n-1)-(m-1); n-m;* and *(n+1)-(m+1)*. These are the descriptions of the event that belong to these particular time periods in *SORta*.

At time moment τ_a the description of the trajectory for the time period of

$$\Delta\tau_a = \tau_a - 0 = \tau_a \quad \text{1E6}$$

is a curve that connects points *(n+1)* on axis *y* and *(m+1)* on axis *x*.

Its length is: $L = c\tau_a$ 1E7

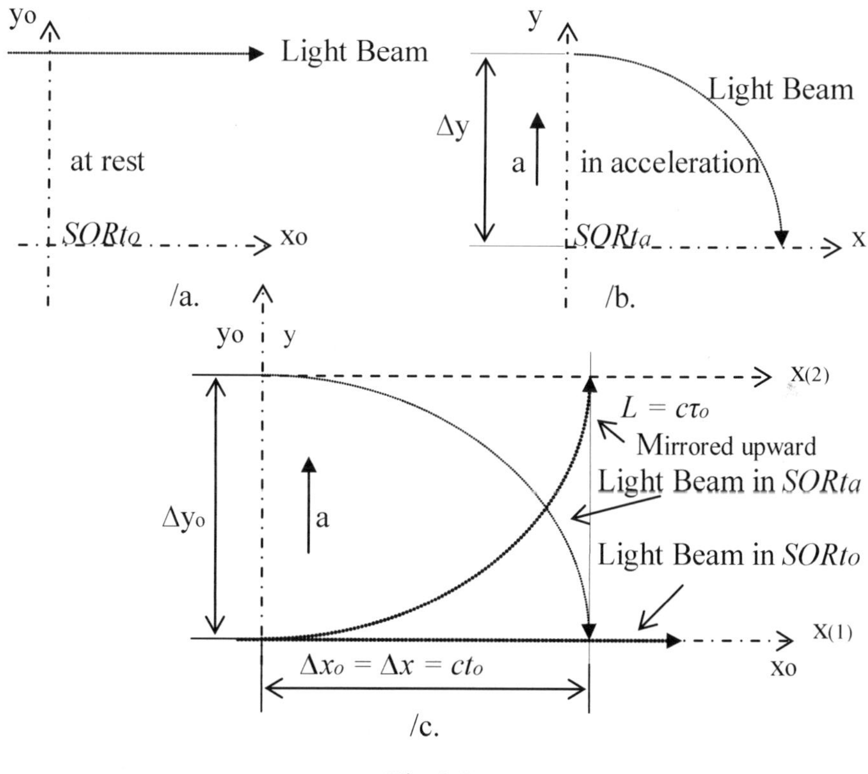

Fig.1.9. Fig 1.9

We use Fig.1.9/c for the assessment of the time relation, with the following inputs:

>> The event happens in *SORto* for time period, measured as $\Delta t_o = t_o - 0 = t_o$ 1F1

>> The description of the event in *SORto* is $\Delta x_o = ct_o$ 1F2

1F3 >> The duration of the event in *SORta* is $\Delta\tau_a = \tau_a - 0 = \tau_a$

>> The description of the light beam in *SORta* is $L = c\Delta\tau_a = c\tau_a$

The motion has no impact on the direction and the velocity of the light. *The light beam still crosses the system of SORta without suffering any impact.* The *time phenomenon* is the only parameter that can be subject to any impact by the accelerating *SORta.*

1G1
1G2
The length of the curve is $L = \int \sqrt{1 + \left(\frac{at_o}{c\sqrt{1 - a^2 t_o^2 / c^2}} \right)^2} \, cdt_o$; and $L = \frac{c^2}{a} \arcsin \frac{at_o}{c}$

L is the length of the curve or the *description* of the light beam in *SORta*: $L = c\tau_a$

1G3 $$\tau_a = \frac{c}{a} \arcsin \frac{at_o}{c}$$

1G3 is the time relation for a particular time period between *SORta,* the accelerating and *SORto*, the system at rest. For a certain time moment it is:

1G4 $$\frac{d\tau_a}{dt_o} = \frac{1}{\sqrt{1 - \left(a^2 t_o^2 / c^2\right)}}$$

The acceleration *does modify* the time flow!
Events in an accelerating system of reference need more time than events in a stationary one. The time in a system of reference in acceleration flows faster.

If the light signal enters *SORta* and crosses the system under angle α, which is of any free value, with reference for the motion with constant speed, the vector components of the light beam will be *equal* to the genuine light signal vector itself:

$$LB_x = LB_y = LB = c\Delta t_o$$

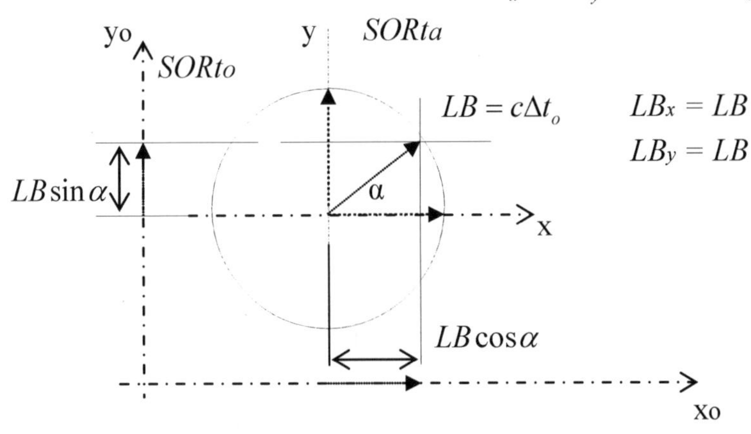

Fig 1.10

Fig.1.10

The time relations for acceleration:

$$\tau_{a(parallel)} = \tau_{a(perpendicular)} = \tau_{a(anydirection)} = \tau_a = \frac{c}{a}\arcsin\frac{at_o}{c}; \text{ and} \qquad \text{1G5}$$

The time relations are independent of the direction of the motion of the system of reference.

$$\boxed{\frac{d\tau_a}{dt_o} = \frac{1}{\sqrt{1-\frac{a^2t_o^2}{c^2}}}} \qquad \text{1G6}$$

(Explanations in full on these subjects are given in Book 1, Sections 1, 2 and 3.)

Part 2 — P.2
Change of the mass status is the key – drive of time
What is the work, necessary for the acceleration of mass system of reference?

We divide the process of the acceleration into an infinite number of equal and *as small as possible* time periods. We are taking the gradual increase of the velocity of *SORta*. The speed grows in time points $t_{o(n-1)}$, $t_{o(n)}$ and $t_{o(n+1)}$. It is constant in each period, but period by period of higher value.

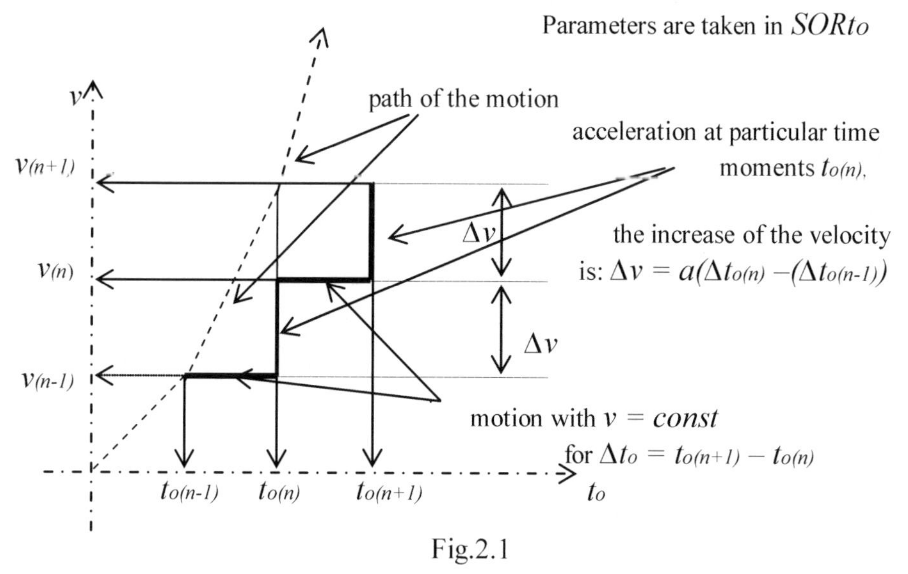

Fig.2.1 — Fig 2.1

What is the value of the work that is necessary to accelerate *SORta*, a system of reference or inert body of (*inert*) mass *m* within *SORto* from speed *0* to *v*?

We are not interested here as to what the energy source of the acceleration of *SORta* is. Whatever it is, the work that makes it happen is to be applied and spent in *SORto*, the stationary system of reference and received by *SORta* in acceleration.

2A1 The work, received by *SORta* for the small as possible time periods of Δto with $v = const$ is: $$dW_a = \frac{dp_a}{d\tau_a} ds_a$$

2A2 The same work provided within *SORto* is: $$dW_o = \frac{dp_o}{dt_o} ds_o$$

where *pa* and *po* are the momentums of *SORta* calculated in *SORta* and *SORto* accordingly; *τa* – is time measurement in *SORta*; *to* – is time measurement in *SORto*; and *sa* and *so* are the paths of the acceleration, *SORta* makes for the smallest possible time period measured and calculated relative to *SORta* and *SORto* accordingly.

Building up the differentials in 2A1 and 2A2 we are getting to:

2A3 $$dW_a = \frac{dp_a}{d\tau_a} ds_a = \frac{d(mv)}{d\tau_a} ds_a = \left(\frac{dm}{d\tau_a} v + \frac{dv}{d\tau_a} m\right) ds_a$$

2A4 $$dW_o = \frac{dp_o}{dt_o} ds_o = \frac{d(mv)}{dt_o} ds_o = \left(\frac{dm}{dt_o} v + \frac{dv}{dt_o} m\right) ds_o$$

We are taking *SORta* is accelerating only at certain time points, and within the smallest possible time periods the speed is $v = const$.

Therefore in 2A3 and 2A4 $\frac{dv}{d\tau_a} = 0$ and $\frac{dv}{dt_o} = 0$.

We keep however the value of the mass as function of the time flow, to be further examined from the point of view of both systems of reference. The growth of the speed is proportional to the duration of the previous, small as possible time periods and equal to $dv = adt_o$; the value of the acceleration is measured in *SORto* and is $a = const$.

2A5 2A3 and 2A4 can be written as: $dW_a = \frac{dm}{d\tau_a} v ds_a$; and $dW_o = \frac{dm}{dt_o} v ds_o$;

The description of the length of the path of the acceleration at a certain time moment *to* is: $$ds_a = ds_o \frac{1}{\sqrt{1 - (a^2 t_o^2 / c^2)}}$$

and the time relation at this time moment is: $$d\tau_a = dt_o \frac{1}{\sqrt{1 - (a^2 t_o^2 / c^2)}}$$

It is important to note: the speed, result of acceleration is measured within *SORto* and it is *reciprocal*!

$$dW_a = \frac{dm}{d\tau_a} vds_o \frac{1}{\sqrt{1-\left(a^2t_o^2/c^2\right)}} \text{; gives } d\left(dW_a\right) = \frac{dm}{d\tau_a} adt_o ds_o \frac{1}{\sqrt{1-\left(a^2t_o^2/c^2\right)}} \quad \text{2A6}$$

$$dW_o = \frac{dm}{d\tau_a \sqrt{1-\left(a^2t_o^2/c^2\right)}} vds_o \text{; gives } d\left(dW_o\right) = \frac{dm}{d\tau_a \sqrt{1-\left(a^2t_o^2/c^2\right)}} adt_o ds_o \quad \text{2A7}$$

2A6 and 2A7 are obviously equal, but the components of the work formula are different:

- The *accelerating force* in 2A6 is:

$$F_a = \frac{dp_a}{d\tau_a} = v\frac{dm}{d\tau_a} \text{ gives } dF_a = adt_o \frac{dm}{d\tau_a} \text{ and } dF_a = a\sqrt{1-\left(a^2t_o^2/c^2\right)}dm$$

- The *accelerating force* in 2A7 is:

$$F_o = \frac{dp_o}{dt_o} = v\frac{dm}{\sqrt{1-\left(a^2t_o^2/c^2\right)}d\tau_a} \text{ gives } F_o = \frac{adt_o dm}{\sqrt{1-\left(a^2t_o^2/c^2\right)}d\tau_a} \text{ and } dF_o = adm$$

$$\text{Consequently} \quad F_a < F_o \quad \text{and} \quad F_o = F_a \frac{1}{\sqrt{1-\left(a^2t_o^2/c^2\right)}} \quad \text{2A8}$$

- *less F_a force acts for longer period $d\tau_a$ and in longer path ds_a.*
- *larger value of F_o force acts for less dt_o time period and in less ds_o path.*

Whereas the speed relation between *SORta* and *SORto* is reciprocal, the time systems are different, the value of the *acceleration, from the point of view of SORta,* the system of reference in acceleration (*and would-be-measured* within it), is permanently changing: $a_{(SORta)} \neq const$;

$$a_{(SORta)} = \frac{dv}{d\tau_a} = \frac{dv}{dt_o}\sqrt{1-\frac{a^2t_o^2}{c^2}}; \quad \text{and} \quad a_{(SORta)} = a\sqrt{1-\frac{a^2t_o^2}{c^2}} \neq const \text{ the}$$

where $a = a_{(SORto)}$

For getting correct results, mass change impulse shall be used, which includes this difference of the change (acceleration).

$$dF_a = \frac{dv}{d\tau_a}dm = \frac{dv}{dt_o}dm\sqrt{1-\frac{a^2t_o^2}{c^2}} = \frac{dv}{dt_o}dm_a; \quad impulse = \frac{dF_a}{dv} = \frac{dm}{d\tau_a} \quad \text{2B1}$$

$$dF_o = \frac{dv}{dt_o}dm \;; \quad impulse = \frac{dF_o}{dv} = \frac{dm}{dt_o} \quad \text{2B2}$$

Approaching the same speed value, the mass values of the *change* are significantly different

$$\frac{dm_a}{dt} = -\frac{dm}{dt}\sqrt{1-\frac{a^2t_o^2}{c^2}} \quad \text{2B3}$$

[The *acting* mass within the system of reference of the event – the system of reference, which is in acceleration – is getting less and less: $m_a = m\sqrt{1-(v^2/c^2)}$]

With substitutions

$dt_o = d\tau_a \sqrt{1-(a^2 t_o^2 / c^2)}$;

$dm_a = dm\sqrt{1-(a^2 t_o^2 / c^2)}$

$dW = dW_a = dW_o$

as $ds_o = vdt_o = at_o dt_o$

$\frac{dW}{dm_a} = C - c^2 \cos\alpha$

taking $C = c^2$

The solution and the forms of 2A6 and 2A7 will be

$$d(dW) = \frac{dm_a}{\sqrt{1-(a^2t_o^2/c^2)}} \frac{\sqrt{1-(a^2t_o^2/c^2)}}{dt_o} a \frac{dt_o ds_o}{\sqrt{1-(a^2t_o^2/c^2)}}$$

and

$$d(dW) = \frac{dm_a}{\sqrt{1-(a^2t_o^2/c^2)}} ads_o = \frac{dm_a}{\sqrt{1-(a^2t_o^2/c^2)}} a^2 tdt_o$$

$$d\frac{dW}{dm_a} = a^2 \int \frac{t_o}{\sqrt{1-(a^2t_o^2/c^2)}} dt_o \text{ ; } \frac{dW}{dm_a} = c^2 - c^2\cos\alpha = c^2\left(1-\sqrt{\sin^2\alpha}\right)$$

2B4 The results of 2A6 and 2A7 *fundamental* and are

$$dW = dW_o = dW_a = dm_a c^2 \left(1 - \sqrt{1 - \frac{a^2 t_o^2}{c^2}}\right)$$

- *Mass change results in* acceleration.
- *Acceleration of mass* for the count if its internal "energy" – *as* universal event – *establishes TIME!*

Mass change in time = event. Event is measured in time. = No event means no time!

From 2B4 follows that: $E = mc^2$ is the absolute energy of the original mass!

[[In the case of $dt_o = d\tau_a$ gives the *Newtonian* formula:]] $W_a = W_o = \frac{1}{2} m_a a^2 t_o^2$

P. 2.1

2.1. Intensities of the mass change

The change of the absolute work value for the stationary system of reference and the system of reference in motion in 2B4 are equal. It corresponds however to different *intensities* for time period in *SORto* $\Delta\tau_a = \tau_a - 0 = \tau_a$ and

time period in *SORta* $\Delta t_o = t_o - 0 = t_o$ respectively:

2C1
$$w_a = \frac{dW_a}{d\tau_a} = \frac{dm_a c^2}{d\tau_a} - \frac{dm_a c^2}{d\tau_a}\sqrt{1-\frac{v^2}{c^2}}$$

2C2
$$w_o = \frac{dW_o}{dt_o} = \frac{dm_a c^2}{dt_o} - \frac{dm_a c^2}{dt_o}\sqrt{1-\frac{v^2}{c^2}} = \frac{dm_a c^2}{d\tau_a\sqrt{1-\frac{v^2}{c^2}}} - \frac{dm_a c^2}{d\tau_a}$$

2C1 and 2C2 are the real *appearances* of the work within these systems of reference. *Work intensities* are not equal: $w_a \neq w_o$

2C1 and 2C2 have different meaning:
2C1 – acceleration for the count of the internal energy: loss in mass
2C2 – acceleration for the count of external energy: mass is constant, but the intensity of the work of the external acceleration is increased.

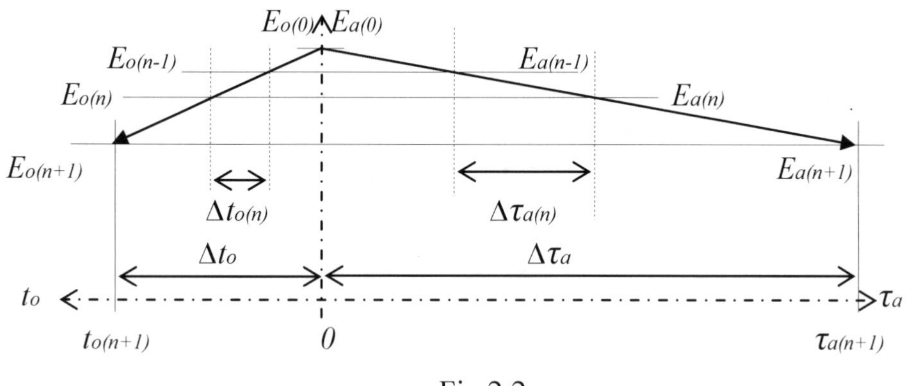

Fig.2.2

Fig 2.2

Absolute values are equal:

$$W_o = E_{o(0)} - E_{o(n+1)} = W = E_{a(0)} - E_{a(n+1)} = W_a$$

Intensities are different:

$$w_o = \frac{dW_o}{dt_o} \neq \frac{dW_a}{d\tau_a} = w_a \qquad \text{2C3}$$

$$\frac{dW_o}{dt_o \varepsilon_o} = \frac{dW_a}{d\tau_a \varepsilon_a} \text{ is equivalent to } \frac{w_o}{\varepsilon_o} = \frac{w_a}{\varepsilon_a}; \text{ since in general meaning: } \varepsilon = \frac{1}{dt}$$

If we take the intensity of the event in the supposed system of reference at rest as $\varepsilon_o = 1$,

$z_a = \varepsilon_a / \varepsilon_o$ will denote the *event concentration* of the system of reference of the acceleration relative to the system of reference at rest.

The *intensity of events* within a system of reference relates to all physical indicators of the occurrence. The *intensity of work* necessitates introducing category *e,* the *intensity of energy* as well. It characterises, in fact, the intensity of the working capability (of the event) in the circumstances of the given system of reference:

$$\Delta e = w_{(n)} - w_{(n-1)} \quad or \quad e_{(n+1)} = e_{(n)} + w \qquad \text{2D1}$$

$$\frac{dW}{dt_o} = \varepsilon_o = w_o \text{ and } \frac{dE}{dt_o} = \varepsilon_o = e_o; \qquad \frac{dW}{d\tau_v} = \varepsilon_v = w_v \text{ and } \frac{dE}{d\tau_v} = \varepsilon_v = e_v \qquad \text{2D2}$$

$$\frac{dW}{d\tau_a} = \varepsilon_a = w_a \quad \text{and} \quad \frac{dE}{d\tau_a} = \varepsilon_a = e_a \qquad \text{2D3}$$

z, the event concentration is *inversely proportional* to the time relations, therefore

2D4 $$z_x = \frac{\varepsilon_x}{\varepsilon_o} = \frac{dt_o}{d\tau_x}; \quad \varepsilon_x = \varepsilon_o z_x \quad \text{and} \quad e_x = z_x e_o; \quad \text{or} \quad w_x = z_x w_o$$

where w_x, e_x, z_x and w_o, e_o denote *work* and *energy intensities* and *event concentration* values within the system of reference in motion and at rest.

We have also found in 2C1, 2C2 the potential of a mass transformation and re-transformation: Should the mass at the start of the (external) process (2C2) be equal to the end stage of the "acting mass" in transformation (internal) process:

2D5 $$w_a = \frac{dm_a c^2}{d\tau_a} - \frac{dm_a c^2}{d\tau_a}\sqrt{1-\frac{v^2}{c^2}};$$

the re-transformation is full

2D6 $$w_o = \frac{dm_a c^2}{d\tau_a}\sqrt{1-\frac{v^2}{c^2}} - \frac{dm_a c^2}{d\tau_a}\frac{\sqrt{1-\frac{v^2}{c^2}}}{\sqrt{1-\frac{v^2}{c^2}}} = = \frac{dmc^2}{dt\tau_a}\sqrt{1-\frac{v^2}{c^2}} - \frac{dm_a c^2}{d\tau_a}$$

(Explanations in full on these subjects are given in Book 1, Sections 5 and 6.)

P.3

Part 3
Energy quantum
is the non-mass status of the matter in transformation

Mass transformation can be described as

3A1 $$d\left(m_o c^2 - m_o c^2\sqrt{1-\frac{v^2}{c^2}}\right) = \frac{dp_{tr}}{dt_{tr}} ds_{tr}; \quad \text{and} \quad d\left(m_o c^2 - m_v c^2\right) = \frac{d(m_{tr}\upsilon)}{dt_{tr}} ds_{tr}$$

where

dp_{tr} – is the momentum, *result* of the transformation of the mass;

dt_{tr} and ds_{tr} – are the duration and the path of the mass transformation;

m_{tr} – is the value of the transformed mass; m_v is the speeded up value of m_o; and

υ – is the velocity of the mass transformation, the speed value of the "disappearance" of the matter. It is not equal to v or Δv, the speed or the speed difference between the systems of reference, the different levels of the sphere symmetrical expanding acceleration of the matter.

The formula above means, the difference in mass values is transforming into energy. The energy is the matter in motion at speed υ.

Can the value of υ in 3A1 be other than c?

$$d\left(m_o c^2 - m_v c^2\right) = \left(\frac{dm_{tr}}{dt_{tr}}\upsilon + \frac{d\upsilon}{dt_{tr}} m_{tr}\right)\upsilon dt_{tr} \qquad \text{3A2}$$

For simplicity we take the shortest as possible time period and suppose that the speed of the mass transformation for this period is constant in time:

$$\frac{d\upsilon}{dt_{tr}} = 0$$ (The result will prove that this condition is taken correctly.)

3A2 gives:

$$\frac{d\left(m_o c^2 - m_o c^2 \sqrt{1 - \frac{v^2}{c^2}}\right)}{dm_{tr}} = \upsilon^2 \qquad \text{3A3}$$

the integral of which results in:

$$m_{tr} = m_o \frac{c^2}{\upsilon^2}\left(1 - \sqrt{1 - \frac{v^2}{c^2}}\right) \qquad \text{3A4}$$

3A3 is the evidence: *there is only one valid option:* $\upsilon = c$

With reference to 3A3, the value of υ, the speed of the "disappearance" of the mass into energy (motion) *cannot be either more or less than c*, otherwise the mass balance cannot be maintained:

$$m_{tr} c^2 = m_o c^2 - m_o c^2 \sqrt{1 - \frac{v^2}{c^2}} \qquad \text{3A5}$$

We have to note here that the speed of the system of reference in motion, is still $v < c$, equal to the speed of one of the levels of the step by step sphere symmetrical expanding acceleration of the mass.

There are here two kinds of transformation:

- the acceleration of the system of reference from one status with certain speed and certain mass, into another with higher speed and less mass; and
- the transformation of the mass into energy with always constant speed of the "mass disappearance" as $\upsilon = c = const$,

At $v = c$, the mass is zero, but the energy exists:
the equal energy quantum ($e = q$) – as result of mass transformation
– *has been born.*

Mass transformation directly produces *energy*, creating the *Quantum System of Reference*, a homogenous and stable-in-speed energy field, with an infinite number of *energy quantum*. The *mass system of reference* in *transformation* is approaching the *Quantum System of Reference* with less and less mass values.

Were there any energy exchange between energy quantum, the work formula of the exchange, should have to be written as

3B1
$$\frac{de}{dq} = \left(1 - \sqrt{1 - \frac{(c_{q1} - c_{q2})^2}{c^2}}\right)$$

It would mean a speed difference, value of $(c_{q1} - c_{q2})$, between quantum.

We may, of course, turn this statement on its head, saying that should there be any speed difference this would result in energy exchange between them. Rearrange 3B1 the resulting expression is

3B2
$$\frac{de}{\sqrt{1 - \frac{|c_{q1} - c_{q2}|^2}{c^2}}} = \frac{dq}{\sqrt{1 - \frac{|c_{q1} - c_{q2}|^2}{c^2}}} - dq;$$

3B3 $q = const$ means $c = c_{q1} = c_{q2}$

(Explanations in full on the subjects are given in Book 2, Sections 14 and 15.)

P.4

Part 4
Sphere symmetrical expanding acceleration for infinite time

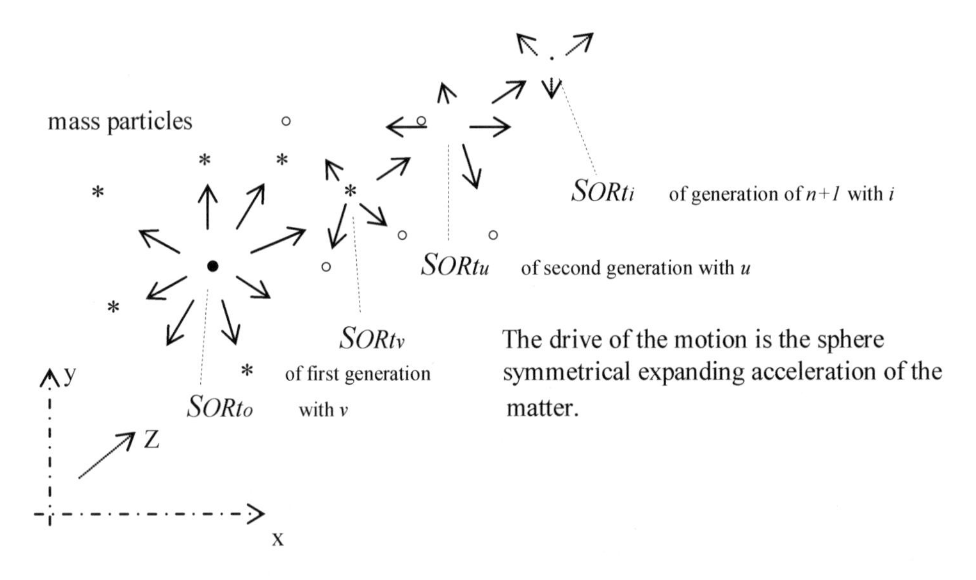

Fig 4.1

Fig. 4.1

As *mass change* is establishing *time*, what is (are) the form/s of the mass change, which last for infinity?

We examine *sphere symmetrical expanding acceleration* (and later collapse) of mass for infinite time. This motion shall have no centre at rest.

Once the expanding acceleration starts, there will be an infinite number of mass components and particles in motion. Each and all of them can be considered as an origin of further mass expansion, with generation of new mass particles. Since the energy of mass is transforming into energy of motion, the mass of these particles in sphere symmetrical expanding acceleration is expected to be always less than the mass of their origin.

Fig 4.1 is a simplified projection of the sphere symmetrical expanding acceleration of the matter from its stationary *SORto* status, through an infinite number of intermediate stages, finally into motion with $i = \lim a\Delta t = c$, *SORti*. Systems of reference *SORtv* and *SORtu* represent the acceleration of the mass components at actual speed v and u. Together with *SORti* they mark different levels of the motion of mass particles.

Why sphere symmetrical expanding acceleration?
- Because any other possibility of the transformation can either be excluded or covered:

➢ motion with $\frac{dv}{dt} = 0$ is not excluded, but this is not about mass transfer, rather status of $\frac{de}{dm} = 0$, the constant energy; this can hardly occur until the full energy of the mass has been exhausted.

➢ Until $\frac{dv}{dt} > 0$ is valid, the direction of the acceleration of mass (particles) in fact is irrelevant. Sphere symmetrical expanding acceleration represents any direction perpendicular to the supposed spherical surface of the mass components in transformation.

➢ $\frac{dv}{dt} < 0$ is about sphere symmetrical accelerating collapse – event in the opposite direction.

➢ Collision of mass particles with each other is an important factor, but in this case it is irrelevant, and does not influence in substance the results of the transformation.

➢ We could also imagine a "rocket type" acceleration of mass components of matter, which might accelerate mass particles in a variety of random directions. The need to use the internal energy of the mass, the transformation of the mass into motion, for making the acceleration to happen, would have, however, been the same. "Rocket type" acceleration of mass components in any direction without centre at rest, is in fact identical to sphere symmetrical expanding acceleration.

We assume that the necessary internal "energy" of mass is granted to accelerate it up to $v = i = \lim a\Delta t = c$, up to quasi the "speed of light".

Acceleration – as event – is part of the existence of matter. The infinity of matter can only be proved, if the acceleration is also infinite.

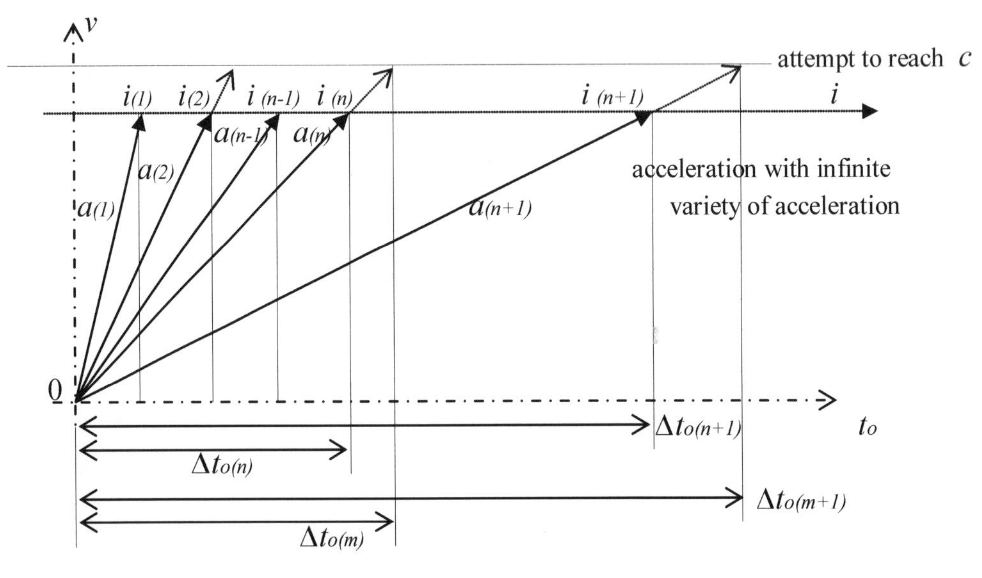

Fig. 4.2

Fig.4.2

Once the mass in acceleration reaches $v = i = \lim a\Delta t = c$, the transformation process will go in balanced way, result of the interaction of the mass system of reference in acceleration and the *Quantum System of Reference*. (Later clearly specified.)

The intensity of work at the *time moment* when reaching $v = i = \lim a\Delta t = c$ is:

- from the point of view of the system of reference in acceleration is:

- from the point of view of the stationary system of reference is:

4A1
4A2

$$w_a = \frac{dm_a c^2}{dt_a}\left(1-\sqrt{1-\frac{i^2}{c^2}}\right)$$

$$w_o = \frac{dm_a c^2}{dt_o}\left(1-\sqrt{1-\frac{i^2}{c^2}}\right) = \frac{dm_a c^2}{dt_a\sqrt{1-\frac{i^2}{c^2}}} - \frac{dm_a c^2}{dt_a}$$

The motion at and above $i = \lim a\Delta t = c$ is permanent collision with quantum particles. The acceleration is permanent and the resistance of the *Quantum System of Reference* is also permanent. It results in constant *try-and-drop-dawn* process – a balanced status – the speed of mass acceleration stays constant.

Motion of mass components is characterised by $a_{(n)}$ acceleration and $\Delta t_{o(n)}$ time period drops down to the next "acceleration line" with $a_{(n+1)}$ and $\Delta t_{o(n+1)}$.

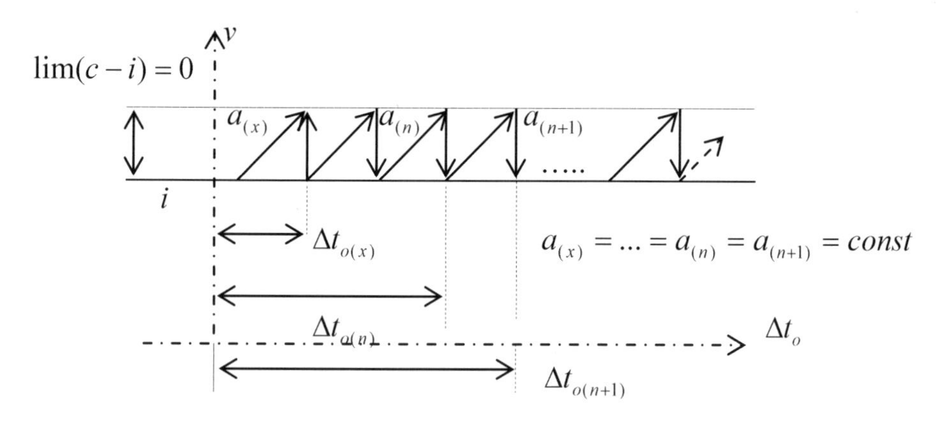

Fig.4.3

Fig. 4.3

or

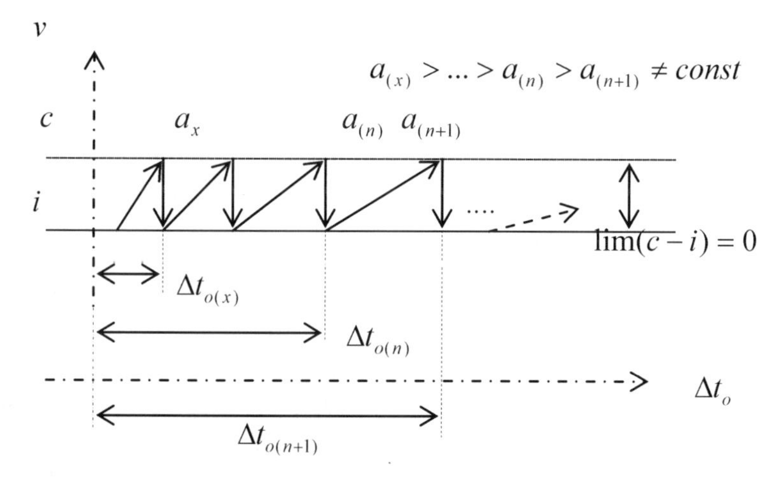

Fig.4.4

Fig 4.4

The work intensity, which is to restore the speed drop-down in each case, must match the energy intensity of the restoration. In the case of *drop-down work* when restoring and reaching $v = i = \lim a\Delta t = c$ again,

$$w_{a(drop-down)} = \frac{dmc^2\sqrt{1-(i^2/c^2)}}{dt_a}\left(1-\sqrt{1-\frac{a_{(n)}^2\Delta t_o^2}{c^2}}\right) = \frac{dm_i c^2}{dt_a}\left(1-\sqrt{1-\frac{(c-i)^2}{c^2}}\right) \quad \text{4A3}$$

the intensity of the work from the point of view of the stationary system of reference is:

4A4
$$w_{o(drop-down)} = \frac{dmc^2\sqrt{1-(i^2/c^2)}}{dt_a\sqrt{1-\frac{a_{(n)}^2\Delta t_o^2}{c^2}}} - \frac{dmc^2\sqrt{1-(i^2/c^2)}}{dt_a} = \frac{dm_ic^2}{dt_a\sqrt{1-\frac{(c-i)^2}{c^2}}} - \frac{dm_ic^2}{dt_a}$$

P.
4.1

4.1.
Gravitation

Gravitation is the *sphere symmetrical expanding acceleration of the Earth*, the system of reference *in motion with* $i = \lim a\Delta t = c$ for infinite time.

The *sphere symmetrical expanding acceleration of the Earth, a motion with* $i = \lim a\Delta t = c$ obviously *modifies* our existing view on the *gravitational free fall.*

The sphere symmetrical expanding acceleration of the *Earth*, the motion with $i = \lim a\Delta t = c$ *for infinite time,* provides the energy for the *blue shift* of electromagnetic waves, reference to Section 9. *It must also provide energy to any other inert body and system of reference* in *"free fall"*. What is the free fall, then?

Ref.
S.9

The *free fall on the Earth is a motion of an inert body or system of reference with speed* $i = \lim a\Delta t = c$ *but without the accelerating component.* The surface of the *Earth* in motion with $i = \lim a\Delta t = c$ *accelerates towards the inert bodies or systems of reference in free fall.*

The collision of electromagnetic waves with the accelerating surface of the *Earth*, the *blue shift*, increases the energy of the electromagnetic waves. The origin of this energy is the sphere symmetrical expanding acceleration of the *Earth*, its motion with $i = \lim a\Delta t = c$. The energy surplus or deficit of any shift can be

4B1 written as
$$\Delta E = E_\gamma\left(1-\sqrt{1-\frac{g^2\Delta t_o^2}{c^2}}\right); \quad \text{or} \quad \Delta E = xE_\gamma \cdot 10^{-k}$$

but, *most importantly*, the kinetic energy of the *Earth* can be transformed through *blue and red* shift sequence *into energy to be used*:

4B2
$$\Delta E_{(red-blue)} = \Delta E_{(red)} - \Delta E_{(blue)} = E_\gamma \frac{g^2\Delta t^2}{c^2} \cdot \frac{1-\sqrt{1-\left(g^2\Delta t^2/c^2\right)}}{1+\sqrt{1-\left(g^2\Delta t^2/c^2\right)}} > 0$$

This energy benefit is *not "an energy from nothing" rather* an energy from *gravitation* to be used.

4.2. Work against the *Quantum System of Reference*

P. 4.2

The sphere symmetrical expanding acceleration is stopped by the energy quantum of the *Quantum System of Reference* at $i = \lim a\Delta t = c = const$. The energy is "taken off" by equal energy quantum in balance with the loss of the mass of the accelerating system of reference.

The loss of the accelerating energy of the mass in motion with $i = \lim a\Delta t = c$ in collision with the *Quantum System of Reference* is equal to:

$$\frac{de}{dt} = \frac{dmc^2}{dt}\sqrt{1-\frac{i^2}{c^2}}\left(1-\sqrt{1-\frac{(c-i)^2}{c^2}}\right) \qquad \text{4C1}$$

dm – mass of the system of reference in sphere symmetrical expanding acceleration, on the approach to $v = c$, in motion with quasi constant $i = \lim a\Delta t = c$;

de – the energy, which would accelerate this particular mass from speed i up to speed c;

$(c - i)$ – is the speed difference for reaching the quantum stage.

In accelerating terms: $adt = dv = (c - i)$

4.3. *Quantum Entropy*

P. 4.3

The acceleration at speed $i = \lim a\Delta t = c$ continues but the general rule of entropy and the resistance of the *Quantum System of Reference* prevents transforming the entire mass into motion. We can call this *quantum entropy* or *entropy of the mass transformation* at $i = \lim a\Delta t = c$:

$$s_i = \dot{m}_i c^2\sqrt{1-\frac{(c-i)^2}{c^2}} = e_{qi} \qquad \text{where: } \dot{m}_i = \frac{dm_i}{dt_i} \qquad \text{4D1}$$

$$\text{since} \quad \frac{dmc^2}{dt}\sqrt{1-\frac{i^2}{c^2}}\left(1-\sqrt{1-\frac{(c-i)^2}{c^2}}\right) < q_{qi} \qquad \text{4D2}$$

The value of the quantum entropy cannot be decreased: the energy of the mass of the system of reference of the matter in motion with $\lim i = c$ is not sufficient for any further transformation.

(Full explanations on the subjects are given in Sections 7, 8, 9, and 10 of Book 1 and Sections 14, 15 of Book 2.)

P.5

Part 5
Sphere symmetrical accelerating collapse

With reference to 4C2, there will be a point in the transformation of *matter* from its mass status into energy, when the internal energy of mass of

5A1
$$mc^2\sqrt{1-\frac{i^2}{c^2}}\sqrt{1-\frac{(c-i)^2}{c^2}}$$

in sphere symmetrical expanding acceleration, will be not sufficient for accelerating any more against the opposing energy of the *Quantum System of Reference*. The internal energy of the mass cannot keep the motion with speed $i = \lim a\Delta t = c$ and the slow-down of the sphere symmetrical expanding acceleration starts. As a direct consequence, the *quantum entropy* increases.

The *Quantum System of Reference*, "external" relative to the mass system of reference slows down the acceleration:

5A2
$$\dot{m}_i c^2\sqrt{1-\frac{(c-i)^2}{c^2}} - \frac{\dot{m}_i c^2\sqrt{1-\frac{(c-i)^2}{c^2}}}{\sqrt{1-\frac{(c-v)^2}{c^2}}}$$
- the collapse is result of external impact

As a consequence of the slowing down, the *quantum entropy* grows.
The sphere *symmetrical accelerating collapse* starts.

With the sphere symmetrical expanding acceleration and with the sphere symmetrical accelerating collapse we are getting to a certain "pulsation" of *matter*, which leads to the two ends of its existence.

5A3 At one end to the *quantum entropy*, equal to $s_i = \dot{m}_i c^2\sqrt{1-\frac{(c-i)^2}{c^2}} = e_{qi} = f_i \cdot q$

and at the other, with reference the "*mass entropy*", equal to

5A4
$$e_v = \frac{\dot{m}_i c^2\sqrt{1-\frac{(c-i)^2}{c^2}}}{\sqrt{1-\frac{(c-v)^2}{c^2}}} = s_v = f_v \cdot q$$

During this pulsation, the matter in its mass form cannot reach in its full extent neither the *Quantum System of Reference* nor the *system of reference of absolute rest.*

No event within the stationary system of absolute rest means *no time.* No frequency would be identified, since no-event (no-motion) could not result in any quantum impact.

Part 6
Energy Quantum

P.6

Each *energy quantum* of the *Quantum System of Reference* is product of mass transformation. They keep the balance with mass systems of reference in sphere symmetrical expanding acceleration. Their balancing number may grow, but the energy of each energy quantum stays constant:

$$q = const$$

The energy of each (photon=) *energy quantum* is not just constant but with reference to 3B1 and 3B2, also must be equal to each other. Why? Because

- should we find a single *energy quantum* with higher energy than the energy of the others, the smallest existing, it would mean that the transformation of a certain mass particle of the matter resulted in extra energy, and there are energy quantum around with higher energy than the energy of the mass particles of the matter still in acceleration.
- should we find a single energy quantum with less energy than the others, it would mean that the transformation of the matter into energy ended at a lower quantum energy level and there are mass particles around with less energy than the energy quantum.

Neither of the two cases above can be valid.
(Photon=) *energy quantum* must be of equal energy the smallest possible.

What is the work of the slowing down effect in balance with the sphere symmetrical expanding acceleration of mass systems of reference for infinite time, when energy quantum does not participate in energy exchange?
Max Planck has given *H* the Planck-constant the correct dimension.

$$dmc^2\sqrt{1-\frac{i^2}{c^2}}\left(1-\sqrt{1-\frac{(c-i)^2}{c^2}}\right)=\frac{dn}{dt_i}q\cdot dt_i=f_i(q\cdot dt_i)=f_i\cdot H \qquad \text{6A7}$$

For the intensity circumstances of the *Earth*, it is $(q\cdot dt_i)=H=6.626\cdot 10^{-36}$ [*Joule*·**sec**]

At frequency $f=1$, $(q\cdot dt_i)$ provides the energy of a single quantum:

$$q=(q\cdot dt_i)\frac{1}{dt_i}=q \qquad \text{since} \left[f=\frac{1}{dt_i}\right]$$

The *Planck* constant perfectly matches the experimental results.

It is valid however (only) to a certain acceleration (intensity) value – the system of reference of *Earth*.

P. 6.1

6.1
Energy quantum, the smallest energy portion of the nature is impacted by events

For practical reasons we use $\Delta v = a\Delta t = c - i$ for denoting the speed difference between the system of reference in motion with $i = \lim a\Delta t = c$ and the *Quantum System of Reference*. $(c - i)$ is an infinitely small and constant speed difference of the acceleration. We also use $(c - v)$ in quantum entropy formulas for the characterization of the speed difference, result of the sphere symmetrical collapse.

Both $(c - i)$ and $(c - v)$ are real speed differences relative to c and always relate to the system of reference of the mass status of the matter in motion.

In spite of the examples above, any speed deduction from or summarization to the speed of energy quantum gives identical results:

6B1 $$c + v = c \quad \text{and}$$

6B2 $$c - v = c$$

The collision of a system of reference of measured mass in motion with v with an *energy quantum* of speed c results in speed c of the *energy quantum*. Do 6B1 and 6B2 mean in this case the end of the mathematical correctness of summarizing speed values? No!

On the contrary, they are the drives for finding the correct explanation:
Energy quantum is not accelerated up and is not slowed down as a result of collision with systems of reference of mass in motion (event). Systems of reference in collision with quantum particles neither add nor take off energy from them. The energy quantum is and remains in any circumstances of *equal* and *constant*.

= Energy quantum particles *are the energy themselves!*

The *Quantum System of Reference* is transferring the impact and loaded as *Quantum Membrane*

- The collision of systems of reference in motion with energy quantum can only be characterized by the intensity of the collision.
- The intensity of the collision is the impact of the motion: the number of the energy quantum of the collision. No motion (no event) theoretically means no collision.

- The measurement of the intensity is the *frequency*, the number of the impacted photons in collision for unit period of time. If the frequency before the collision was $f_q = 0$, afterwards it will be a certain value of f_v, depending on the energy intensity of the motion of the system of reference.

The collision with energy quantum
- from in front to the direction of the motion increases the frequency;
- from behind decreases the frequency.

Here we have to note that the frequency cannot be negative. Its value can be less after the collision than before, but cannot be less than *zero*, the original frequency of the photons of the *Quantum System of Reference* without impact.

The energy balance of the matter is about
- the energy of the mass; and
- the energy of the *Quantum System of Reference*.

Energy intensities give the absolute balance if they are corrected by the corresponding event concentration, here in 6C1, in the denominator:

$$\frac{\dot{m}c^2}{1} = \frac{\dot{m}_v c^2}{\sqrt{1-\frac{v^2}{c^2}}} = \frac{\dot{m}_u c^2}{\sqrt{1-\frac{u^2}{c^2}}} = \ldots = \frac{\dot{m}_i c^2}{\sqrt{1-\frac{i^2}{c^2}}}; \qquad \text{6C1}$$

as it corresponds to $$\frac{mc^2}{dt_o z_o} = \frac{mc^2}{dt_v z_v} = \ldots = \frac{mc^2}{dt_i z_i} \qquad \text{6C2}$$

where the index of the mass values denotes the speed of the system of reference of the mass.

A closer look at 6C2 proves that the balance relates not only to the mass, rather to the whole matter, addressing also the photons, the equal energy quantum, the results of the mass transformation into energy.

$$\frac{\dot{m}c^2}{1} = \frac{\dot{m}c^2\sqrt{1-v^2/c^2}}{\sqrt{1-v^2/c^2}} = \frac{\dot{m}c^2\sqrt{1-u^2/c^2}}{\sqrt{1-u^2/c^2}} = \ldots = \frac{\dot{m}c^2\sqrt{1-i^2/c^2}}{\sqrt{1-i^2/c^2}} = mc^2 \qquad \text{6C3}$$

The absolute energy value relates to both forms of the matter.

But where is the energy part?

The energy does not leave the “adiabatic box” with only the mass within it. The energy, the product of the transformation of the matter, the result of the sphere symmetrical expanding acceleration of the mass shall also be within this “adiabatic box”.

What are the coordinates of this "adiabatic box" of the matter?

The overall balance between mass, energy, motion, time count and space coordinates predict the continuity of the matter. It does not allow having space without time count, or time without space, or mass without energy (photons) and (photons) without mass.

Trying to picture any space configuration would mean speculation, but having the matter as infinite as it is, the mass-energy balance and the continuity of the time "flow" guarantee the continuity of the space fully "occupied" with energy (quantum) and mass.

There could not be "empty room" in the "adiabatic box" of the matter.
The boundaries of the existence of the matter (mass and energy in time) are infinite in space : $\lim y_{v \to c} = \infty$
and measured, depending on the motion from $\lim y_{v \to 0} = 0$ to $\lim y_{v \to c} = \infty$

(Full explanations on the subjects in Parts 5 and 6 are given in Sections 15 and 16 of Book 2.)

P.7

Part 7
Particles are the processes of mass-energy transformation

Matter is of *infinite* number of "*mass-component-systems-of-reference*" in transformation. Should all mass components be in the same sequence, all systems of reference of the matter would accelerate and collapse in parallel. In this case all *mass-component- systems-of-reference* would reach the full transformation and full collapse at the same time.

Should however there be differences in the mass transformation, the values of the sphere symmetrical expanding acceleration and collapse there will be a shift between the cycles. This natural shift guarantees that the three statuses of the process are permanently present:

- *(a) sphere symmetrical expanding acceleration of the mass of the matter*, the transformation of the mass into energy. (Quantum energy generation)
- *(b) sphere symmetrical expanding acceleration of the mass for infinite time*, motion with $i = \lim a\Delta t = c$. This is balanced energy status of the expanding mass and the *Quantum System of Reference*. No quantum energy generation.

- *(c) sphere symmetrical accelerating collapse of the mass of the matter*, under the effect of the *Quantum System of Reference* (transformation of energy into mass).

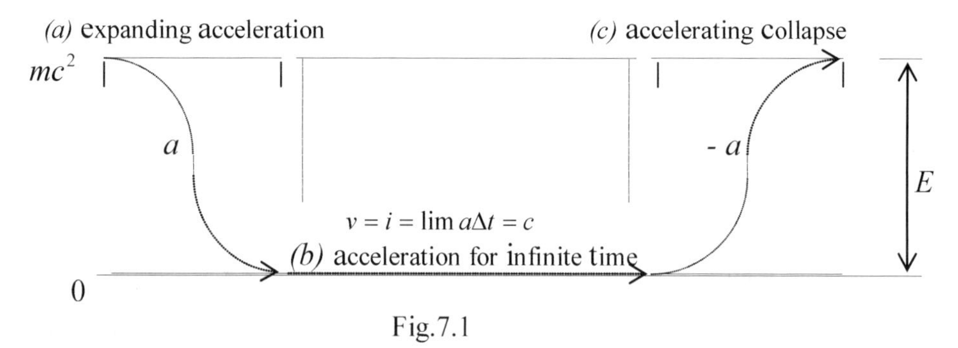

Fig.7.1

Fig 7.1

At the end of the sphere symmetrical expanding acceleration, at the inflection point of the acceleration, the mass components, equal in value to the *quantum entropy* are still in motion with v=lim i= c. ($v < c$) The sphere symmetrical collapse (and slow-down, identified as negative acceleration) starts from this kinetic energy intensity status of matter. At the end of the slow-down (or acceleration relative to the quantum system of reference), at the inflection point, from where the expanding acceleration will start again, the speed of the collapse, is $v > 0$.

7.1 Particles

P. 7.1

The change of mass, acceleration determines the (measurable) presence of the matter, called *elements*.

Proton is the measured effect of the *process* of the sphere symmetrical expanding acceleration, the transformation of mass into energy, denoted through p.

Mass, after reaching speed $i = \lim a\Delta t = c$ continues its sphere symmetrical expanding acceleration for infinite time. This is the *electrons process* denoted through e. *Electrons* lose in mass as a result of keeping the balance with the *Quantum System of Reference*. The process ends at quantum entropy status.

Once the mass change of electrons reaches quantum entropy status, the sphere symmetrical collapse starts. The *neutron* is the process of the sphere symmetrical accelerating collapse, denoted as n. Collapse happens under the effect of the *Quantum Membrane*.

Protons are energy providers, transforming their mass into quantum energy. *Electrons* are "fighters" and keep the balance within the element with the energy of the *Quantum System of Reference*. *Neutrons* are energy catchers. Their collapse re-transforms the quantum energy into mass again. The new cycle starts with less mass.

Speaking about sphere symmetrical expanding acceleration and collapse, our conventional mind may suppose a "central" spot at rest or motion, around which other particles are in expansion or collapse. This view is understandable, but false: there are infinite numbers of spots, *all* in expanding acceleration or accelerating collapse. The acceleration of the mass of the matter cannot be zero.

The definition of the particles is only acceptable if we add: This definition is only for distinguishing purposes on a certain level of our relativistic approach. Otherwise it might suggest we limit the existence of the matter. Protons, electrons, neutrons (and many others) cannot be final restrictive categories. We cannot find "*the particle*". Instead we find the effect of the mass-energy transformation (of the matter) and call it proton, neutron, electron and others. We can count them, but this count is no more than the assessment of the effect – the mass energy balance of the matter (at the level of our measurement and capabilities).

Should there be any spot, central or other, or place at absolute rest within the matter, it would mean no time, no event and, consequently, no matter. Therefore, matter must have an infinite number of systems of reference in motion with $i = \lim a\Delta t = c$, the acceleration for infinite time and in motion at the same time with $i = \lim a\Delta t = c$ relative to each other.

7A1 $$\lim i + \lim i + ... + \lim i = c$$

7A1 means that a slowing down or speeding up from $\lim ii = (\lim i) = c$ to $\lim i = (\lim ii) = c$ is fully acceptable and correct. It means two systems of reference, out of the infinite number, in relative motion to each other by *i*: two "levels" of our relativistic assessments.

P 7.1.1

7.1.1. The mass change of particles

The *proton process* is the change of mass in time

7B1 $$\text{from} \quad m \quad \text{to} \quad m\sqrt{1-\frac{i^2}{c^2}}$$

The *electron process* is mass change

7B2 $$\text{between} \quad m\sqrt{1-\frac{i^2}{c^2}} \quad \text{and} \quad m\sqrt{1-\frac{i^2}{c^2}}\sqrt{1-\frac{(c-i)^2}{c^2}}; \quad \text{and}$$

The *neutron process* is the change of mass

$$\text{from } m\sqrt{1-\frac{i^2}{c^2}}\sqrt{1-\frac{(c-i)^2}{c^2}} \quad \text{to} \quad \frac{m\sqrt{1-\frac{i^2}{c^2}}\sqrt{1-\frac{(c-i)^2}{c^2}}}{\sqrt{1-\frac{(c-v)^2}{c^2}}} \qquad \text{7B3}$$

Since the *particle-processes* are present in the transformation, acceleration and collapse in parallel, the all effects of the change are "measured" together within the total mass of the element: $p\cdot k+e\cdot j+n\cdot r=M$

where M is the atomic weight of the element, *k, j* and *r* denote the total number of protons, electrons and neutrons respectively within the element.

The mass chain is:

$$m \quad m\sqrt{1-\frac{i^2}{c^2}} \quad m\sqrt{1-\frac{i^2}{c^2}}\sqrt{1-\frac{(c-i)^2}{c^2}} \quad m\sqrt{1-\frac{(c-i)^2}{c^2}}\ [=m_{(next)}] \qquad \text{7B1}$$

proton (-process) electron (-process) neutron (-process) [proton of the next cycle]

The balance between the
proton (*sphere symmetrical expanding acceleration*) and
neutron (*sphere symmetrical accelerating collapse*) processes is:

$$\frac{mc^2}{dt_o\varepsilon_p}\left(1-\sqrt{1-\frac{i^2}{c^2}}\right)=\frac{mc^2}{dt_o\varepsilon_n}\sqrt{1-\frac{(c-i)^2}{c^2}}\left(\sqrt{1-\frac{i^2}{c^2}}-1\right) \text{ or in its other form:} \qquad \text{7B2}$$

$$\frac{\dot{m}_p c^2}{\varepsilon_p}\left(1-\sqrt{1-\frac{i^2}{c^2}}\right)=\frac{\dot{m}_n c^2}{\varepsilon_n}\sqrt{1-\frac{(c-i)^2}{c^2}}\left(\sqrt{1-\frac{i^2}{c^2}}-1\right) \qquad \text{7B3}$$

The relation of measured mass change effects between P and N is

$$\frac{N}{P}=\frac{\dot{m}_n}{\dot{m}_p}\sqrt{1-\frac{(c-i)^2}{c^2}} \qquad \text{7C1}$$

The *blue shift* of the electron process is:

$$\frac{dmc^2}{dt_i\varepsilon_e}\sqrt{1-\frac{i^2}{c^2}}\left(1-\sqrt{1-\frac{(c-i)^2}{c^2}}\right)=\frac{dn}{dt_i\varepsilon_e}q=f_{ie}\cdot q \qquad \text{7C2}$$

dt_i corresponds to the speed of the sphere symmetrical expanding acceleration, the motion with $i=\lim a\Delta t=c$.

Measured mass change values at the beginning of the first cycle is: / at the end of the first cycle:

7C3, 7C4
$$\frac{dm}{dt}, \qquad \frac{dm}{dt}\sqrt{1-\frac{(c-i)^2}{c^2}}$$

Quantum energy equivalent (in mass change values) at the beginning: / at the end of the first cycle:

7C5, 7C6
$$0, \qquad \frac{dm}{dt}\left(1-\sqrt{1-\frac{(c-i)^2}{c^2}}\right)$$

Measured mass change the end of the n cycle: / Energy quantum equivalent:

7C7, 7C8
$$\frac{dm}{dt}\left(\sqrt{1-\frac{(c-i)^2}{c^2}}\right)^n \qquad \frac{dm}{dt}\left[1-\left(\sqrt{1-\frac{(c-i)^2}{c^2}}\right)^n\right]$$

7D1 With reference to 7B3 and 7C1 the relation of the measured weight (effect) of neutrons and protons is equal to the relation of the intensities of the two processes:
$$Z_{element}=\frac{N}{P}=\left|\frac{\varepsilon_n}{\varepsilon_p}\right|$$

7D2 It is taken:
$$\varepsilon_e=\frac{\dot{m}_p}{\dot{m}_n}\left(=\frac{dm\cdot dt_o}{dt_o dm}=1\right)$$

7D3
$$\varepsilon_e=\frac{\Delta m}{\Delta t_p}\frac{\Delta t_n}{\Delta m}; \text{ and } \Delta t_p\neq\Delta t_n$$

which obviously corresponds to *1* in an ideal case (and in a fully balanced element it is *1* indeed) but the time shift between the two processes in elements gives to this value real meaning – the intensity difference between the processes

The *proton-electron-neutron* process balance gives:

7D4
$$Z\cdot\varepsilon_e=\sqrt{1-\frac{(c-i)^2}{c^2}}=const \qquad \text{and } \varepsilon_e=\left|\frac{\varepsilon_p}{\varepsilon_n}\right|\sqrt{1-\frac{(c-i)^2}{c^2}}\ *$$

* 7D4 would provide to ε_e non-dimensional character, which is not the case, since ε_e is real physical parameter, but for simplicity we will keep and use the formula with this.

Z is the *event concentration* of elements. It is the quotient of the effects of the *neutron* and the *proton* mass change, equal to the quotient of the intensities of the *neutron* and *proton* processes. Z shows the relative energy demand of the neutron process.

The less the value of the event concentration, the more is the internal energy reserve of the element. At $Z=1$ the intensities of the neutron and the proton processes are equal.

The less the value of Z of the element is, the higher is ε_e, the intensity of the electron *blue shift*. This intensity "increase" comes from the time shift difference between the neutron and proton processes of elements. The higher the relative *blue shift* "reserve" is, the more "aggressive" is the element in reaction with others. It characterises the relation of the *intensity of the energy generation* (by proton) and the *intensity of the energy use* (by neutron) of the element.

Diagram 7.1 shows the values of Z, the event concentration for all elements of the Periodic Table. The horizontal axis is the periodic number. The diagram shows divergences in some places and at the end of the curve. The reason for this could be the preciseness of the measurement and the forecast of the atomic weight at the end of the periodic table.

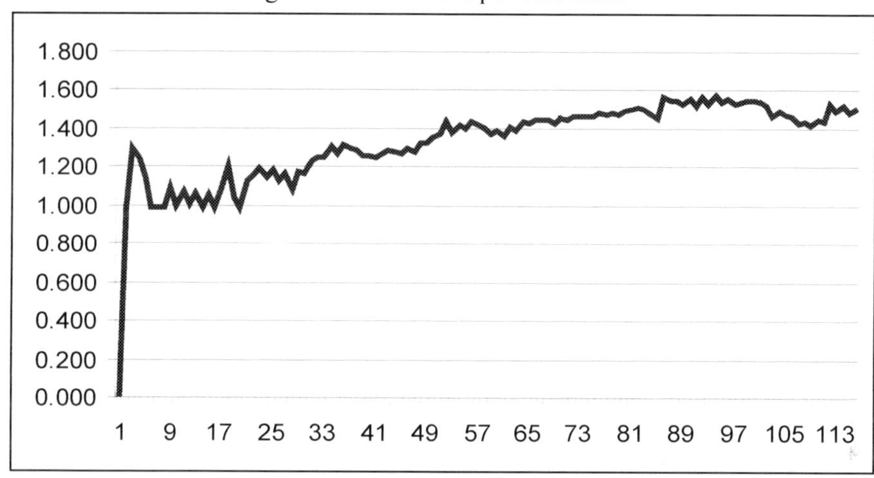

Diag.7.1

Diag. 7.1

M – is the atomic weight of the element, with reference to the periodic table, equal to the weight of neutrons, protons and electrons;

P – is the mass of the protons of the element, calculated from the periodic number and a mass of a single proton;

e – is the mass of the electrons of the atom of the element, calculated from the periodic number and the mass of a single electron;

N – is the mass of the neutrons within the atom of the element calculated as

$$N = M - (P + e)$$

Mass of a single proton is:

1.00727 u or $1.67262171 \cdot 10^{-27} kg$

Mass of a single electron is:

0.00055 u or $9.10938215 \cdot 10^{-31} kg$

where u is the *unified atomic mass*

They are measured within the system of reference of the observation (*Earth*).

The effects of the neutron and proton processes give work value dimension.

P. 7.1.2

7.1.2 ($^{1}_{1}H$) Hydrogen in this chain is *unique*.

The infinite low Z means the duration of the neutron collapse of the *Hydrogen* is infinitely long. The *blue shift* need of the neutron collapse is infinite low. The infinite low intensity of the neutron process – the infinite long time shift between the proton and neutron processes – makes it possible to use the available *blue shift* of the *Hydrogen* for the *red shift* needs of the neutron collapse of other elements.

P 7.1.3

7.1.3 Seven other "specific" elements

Oxygen ($^{8}_{15.9}O$), *Nitrogen* ($^{7}_{14.0}N$), *Helium* ($^{2}_{4}He$), *Carbon* ($^{6}_{12}C$), *Sulfur* ($^{16}_{32}S$), *Calcium* ($^{20}_{40}Ca$) and *Silicon* ($^{14}_{28}Si$).

The $Z<1$ of these elements means: the intensity of the energy generation of the protons is more than the intensity of the energy utilization of the neutrons. Together with the *Hydrogen*, these are the *most* active (chemical reaction) most energy efficient elements: *blue shift* "energy" *providers.*

Elements with $Z<1$ are keen to step into reaction with others and use their *blue shift* intensity "surplus" for intensifying the neutron process of other elements with $Z>1$. These are the most active elements to enter energy relations with others.

The external *blue shift* impact of elements with $Z<1$ (most importantly *Oxygen, Nitrogen, Helium, Carbon*) may result in *blue shift* conflict with elements or compounds with balanced mass-energy structure.

As a general rule, the less *Z*, the event concentration, is, the more energy efficient the element is. The most energy efficient 16 elements are listed in Table 7.1. The values are based on recorded measured mass data, sensitivity of $10^{-27}\,kg$ for the *protons,* and $10^{-31}\,kg$ for the *electrons*. The table also contains those elements, the event concentration of which are close to $Z \geq 1$.

Element	PN	*M*	*Z*
Hydrogen	1	1.00790	0.000081 (!!)
Oxygen	8	15.99900	0.984897
Nitrogen	7	14.00670	0.985971
Helium	2	4.00260	0.986312
Carbon	6	12.01100	0.986841
Sulphur	16	32.06000	0.988744
Calcium	20	40.08000	0.988992
Silicon	14	28.08550	0.991084

Neon	10	20.17000	1.001898
Magnesium	12	24.30500	1.010254
Kalium	19	39.09800	1.042393
Phosphorus	15	30.97370	1.049466
Aluminium	13	26.98150	1.059976
Chlorine	17	35.45300	1.069875
Natrium	11	22.98900	1.074281
Nickel	28	58.71000	1.081108
Fluorine	9	18.99840	1.095154

Table 7.1

Table 7.1

(!!) is a note – data is based on the Periodic Table – not representing the real case, but should be infinite small!

7.2 Unbalanced elementary structures

P. 7.2

With reference to the mass-energy balance of the *proton-electron-neutron process* in general, isotopes, magnets, and all other particles are results of the deficiencies of this balance.

Key factors of the energy balance are: the *event concentration* and the *intensity relations* of the proton-electron-neutron process. These are constant and standard for each element:

$$\varepsilon_e = \frac{\varepsilon_p}{\varepsilon_n}\sqrt{1 - \frac{(c-i)^2}{c^2}} = const \quad \text{and} \quad Z = \frac{N}{P} = \left|\frac{\varepsilon_n}{\varepsilon_p}\right| = const \qquad \text{7E1}$$

P. 7.2.1

7.2.1. Beta decay and gamma radiation

If the proton-neutron process is unbalanced, it will result either in extra or missing electrons within the element.

In the case of *extra electron* unbalance, the extra *blue shift* of the electron surplus must be released, resulting in so-called β *decay*.

In the case of *missing electron* unbalance, the neutron process becomes unstable and the collapse can partially stop and turn into expanding acceleration.

With reference to 19E1, the electron *blue shift* provides energy of the neutron collapse:

7E2 $$\frac{dmc^2}{dt_i}\sqrt{1-\frac{i^2}{c^2}}\left(1-\sqrt{1-\frac{(c-i)^2}{c^2}}\right)=\frac{dmc^2}{dt_i}\frac{\sqrt{1-\frac{i^2}{c^2}}\sqrt{1-\frac{(c-i)^2}{c^2}}}{\sqrt{1-\frac{(i-v)^2}{c^2}}}-\frac{dmc^2}{dt_i}\sqrt{1-\frac{i^2}{c^2}}\sqrt{1-\frac{(c-i)^2}{c^2}} \quad \text{at } v=i$$

The frequency of the collision of the accelerating (instead of collapsing) neutrons with the *Quantum Membrane* depends on the actual speed of the collapse (between $v = \lim 0$ and $v = i$). The lower is the speed the more energy quantum is impacted in the collision for the unit period of time. The frequency will be *blue shifted* rather than *red shifted* by the *accelerating* (instead of collapsing) neutron. It will result in γ decay – high frequency impact to the *Quantum Membrane*.

P. 7.2.2.

7.2.2. Alpha radiation

Should the proton-electron-neutron processes result in mass unbalance, which needs mass release, the element will be safe from destruction with an α - decay.

The *alpha decay* is the release of the smallest possible balanced mass from the structure of the element.
The smallest existing element is the *Hydrogen*, but the intensity of the neutron process of the *Hydrogen* is infinitely low and the process itself is infinitely long. Therefore, ${}_{1}^{2}H$, an element as such, lasts for infinity. Correction cannot be made by releasing a mass corresponding to a value with infinite long process. The smallest mass-energy *sorting-out-portion* of the imbalance is the nucleus of the atom of the ${}_{2}^{4}He$ element.

P. 7.2.3.

7.2.3. Neutron radiation

The fission of the ${}_{92}^{235}U$ *isotope* is result of the increased intensity of their neutron collapse. In natural circumstances proton electron and neutron processes keep balance within the *Uranium* element. In the ${}_{92}^{235}U$ *isotope* the process is unbalanced: the event concentration of the *isotope* is less than the *Uranium* element ($Z_{U235} < Z_{U238}$) and consequently the intensity of the electron process of the *isotope* is more than the intensity of the *Uranium* element ($\varepsilon_{U235}^{e} > \varepsilon_{U238}^{e}$). As a result of the destruction of the *isotope*, the *fission-product-new-elements* remain without sufficient energy (proton process) provision.

They will seek for energy (mass transformation) available in the surrounding environment.

P. 7.2.4.

7.2.4. Mass, with unbalanced energy structure is: magnet

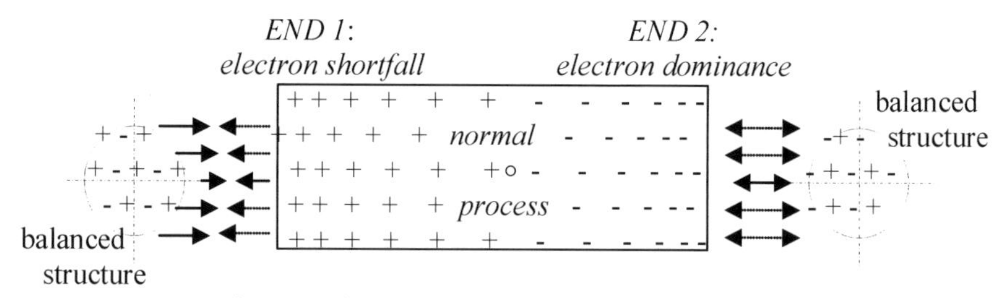

preferred movement *of both subjects towards each other supported by the* blue-red shift sequence *of the sphere symmetrical acceleration of electrons and the collapse of neutrons*

movement *of the subjects towards each other* is hampered *by the* blue shifts *of the sphere symmetrical expanding acceleration of electrons of the elements of both subjects*

Fig.7.2

Fig. 7.2

There is a *blue shift* conflict between the electrons of the electron dominant *END 2* of the magnet and the electrons of the subject on the approach to this *End.*

The electrons in both, the magnet and the subject are impacting the *Quantum Membrane* by *blue shift*. There is a difference between the two impacts. The *blue shift* of the electrons of the normal subject corresponds to normal energy balance conditions. The *blue shift* of the electrons in the magnet at this *End* is more than the *red shift* of the neutron collapse of the magnet at this *End* needs.

The *blue shifts* work against each other and prevent the subject of balanced energy status to approach the magnet at this *End.* The closer is the distance the greater is the thrusting effect between the electrons. The conflict either moves the electrons away from the conflict, or the magnet and the subject on approach will be in antagonistic magnetic thrust.

The missing *blue shift* at the *END 1* to the *red shift* of the neutron collapse is provided by the approaching subject. The magnetic attraction is natural need.

(Full explanations on the subjects are given in Sections 17, 18, 19, 20 and 21 of Book 2.)

II

II

Quantum Engine

S. 24

24

Elementary processes

Ref S.17 S.18

The continuity law of *matter* means balanced mass-energy transformation and re-transformation – in time.
Mass-energy and energy-mass transformations wholly occupy the "space". There is no "space" without matter.

The *proton process* is the transformation of mass into energy. The *neutron process* is the re-transformation of energy into mass. The *electron process* is the drive between the two, the sphere symmetrical expanding acceleration of mass at constant speed of $i = \lim a\Delta t = c$ – impact to the *Quantum System of Reference*, the load of the *Quantum Membrane*.

The proton process (the sphere symmetrical expanding acceleration of mass from $\lim v = 0$ to $i = \lim v = c$) does not load (does not work against it) the *Quantum Membrane*. This expanding acceleration of the mass is – transformation, change the status of matter from mass into energy in time.

Transformation from $\lim v = 0$ to v:

24A1
$$\frac{dm_o c^2}{dt_o}\left(1-\sqrt{1-\frac{\Delta v^2}{c^2}}\right)+\frac{dm_o c^2}{dt_o}\sqrt{1-\frac{\Delta v^2}{c^2}}=\frac{dm_o c^2}{dt_o}; \qquad v-0=\Delta v$$

mass status: $\frac{dm_o}{dt_o}\sqrt{1-\frac{v^2}{c^2}}$; transformation: $\frac{dm_o}{dt_o}\left(1-\sqrt{1-\frac{v^2}{c^2}}\right)$

The transformation is permanent. The mass values and the statuses of the transformation above mean virtual steps of the constant change.

24A1 in other format:

$$\dot{m}_o c^2\left(1-\sqrt{1-\frac{v^2}{c^2}}\right)+\dot{m}_o c^2\sqrt{1-\frac{v^2}{c^2}}=\dot{m}_o c^2 \qquad 24A2$$

which in absolute terms is equivalent to:

$$\frac{dm_o c^2}{dt_o \varepsilon_p}\left(1-\sqrt{1-\frac{\Delta v^2}{c^2}}\right)+\frac{dm_o c^2}{dt_o \varepsilon_p}\sqrt{1-\frac{\Delta v^2}{c^2}}=\frac{dm_o c^2}{dt_o \varepsilon_p} \qquad 24A3$$

and ε_p - is *intensity* of the mass change.

The *intensity* of the mass change is equal to acceleration, the increase of the speed of the mass in transformation. Without change of the speed of the mass there is no event (no time), no acceleration and no mass transformation. Therefore, the acceleration fully characterises the intensity of the mass change.

$$\frac{dm}{dt}\to\frac{dv}{dt}=a\to\varepsilon$$

The dimension of the acceleration is [m/s^2].

The *intensity* characterises the change of equal mass values in time.

Therefore, the dimension of the intensity in 24A3, is taken as [$1/s$].

Intensity values in the formulas of elementary processes not directly relate to the time frame

$$dt\neq\frac{1}{\varepsilon_p}$$

All elements go through all speed values in their proton and neutron processes. All speed values, in both directions of the elementary process (expansion and collapse), correspond to a certain time-frame (directly related to the speed value), the result of the change.

The difference between elements is in the intensities of their proton and neutron processes. The intensity in 24A3 is about this difference, about the process itself: *acceleration*.

The appearance of mass in systems of reference is an effect: its change in time. Generation of energy, as result of mass change, is no other than *the difference in the values* of the *mass change* between two statuses.

Transformation from v_{n+1} to $v=i$:

$$\dot{m}_{n+1}c^2\left(1-\sqrt{1-\frac{\Delta v^2}{c^2}}\right)+\dot{m}_{n+1}c^2\sqrt{1-\frac{\Delta v^2}{c^2}}=\dot{m}_{n+1}c^2; \qquad 24A4$$

$$i=v_{n+1}+\Delta v=\lim v=c \qquad 24A5$$

The transformation of mass into energy results in permanent change of the time systems of the transformation:

Ref S13 24A6

$$dt_{n-1} = \frac{dt_o}{\sqrt{1-(v_{n-1}^2/c^2)}}; \quad dt_{n+1} = \frac{dt_o}{\sqrt{1-(v_{n+1}^2/c^2)}}; \quad dt_i = \frac{dt_o}{\sqrt{1-(i^2/c^2)}}$$

$$dt_o < dt_{n-1} < dt_n < \ldots < dt_i; \qquad \lim dt_i = \propto$$

With reference to Section 13, the various stages of the transformation represent constant balance between mass and energy. The proportions between mass and energy statuses are changing. The total value of the "mass-energy" or "energy-mass" of the matter, expressed through *mass* and *event concentration* values, are equal:

24A7
$$\frac{dm_o c^2}{dt_o \varepsilon_o} = \frac{dm_o c^2}{dt_n \varepsilon_n} = \ldots = \frac{dm_o c^2}{dt_i \varepsilon_i} \quad \text{or} \quad \frac{dm_o c^2}{z_o} = \frac{dm_n c^2}{z_n} = \ldots = \frac{dm_i c^2}{z_i},$$

In 24A7 here, *intensities* strictly relate to the global status, the *end result* of the process. The equality between different statuses is the general rule of the mass-energy balance. (Different to the description of the *intensity* of the acceleration process above.)

The *proton process* has its natural drive: the increased value of the *event concentration* – reciprocal to the value of the time count – changing from $\lim Z_o = \propto$ (end status of the neutron process) to $\lim Z_i = 0$ (start status of the electron process).

The *event concentration potential* of the mass of the *proton* is the drive of the sphere symmetrical expanding acceleration. There is no central spot at rest around which the acceleration of mass happens!

Any central spot at rest would contradict to the continuity law of matter, the "no event means - no time" principle.

There is no conflict and *blue shift* impact between the *proton process* and the *Quantum System of Reference*. (The *conflict* is at the electron *blue shift*, initiating the *"red shift"* of the neutron collapse.)

24A8
$$w_p = \frac{dm_p c^2}{dt_p}\left(1-\sqrt{1-\frac{(a\Delta t)^2}{c^2}}\right); \quad \text{or} \quad w_p = \dot{m}_p c^2\left(1-\sqrt{1-\frac{(a\Delta t)^2}{c^2}}\right)$$

The *continuity law* of the matter only allows "free" mass transformation as proton process, if the neutron process with its collapse "creates" energy gap in the mass-energy balance. The proton gives off mass this way, reduces its internal event concentration potential.

$$\dot{m}_{p(o)} \to \dot{m}_{p(n)} \to \dot{m}_i \quad \text{and} \quad \frac{dm_{p(o)}}{dt_p} > \frac{dm_{p(n)}}{dt_p} > \frac{dm_i}{dt_p}$$

$$\lim v = 0 \to v \to i = \lim v = c$$

$$\lim t = 0 \to t \to \lim t = \propto \quad \text{and} \quad \lim z = 0 \to z \to \lim t = 0$$

$t_o = 1$ is taken in formulas for comparative purposes.

The <u>*electron process*</u> does impact the *Quantum Membrane*.

$$w_e = \frac{dm_o c^2}{dt_i}\left(1-\sqrt{1-\frac{(c-i)^2}{c^2}}\right) = \frac{dn}{dt_i}q \qquad \text{24B1}$$

The mass is changing, the acceleration and the speed increase are permanent but the *electron process* maintains its constant $i = \lim a\Delta t = c$ speed. The *Quantum System of Reference* gains energy from the *blue shift* impact and will be loaded as *Quantum Membrane*. The mass is reaching the status of *quantum entropy*.

$$\frac{dm_o c^2}{dt_o}\sqrt{1-\frac{i^2}{c^2}}\left(1-\sqrt{1-\frac{(c-i)^2}{c^2}}\right) \le e_{qe} \quad \text{and} \quad e_{qe} = \dot{m}_{qe}c^2 \qquad \text{24B2}$$

The mass and the value of the acceleration are changing.

$\frac{c-i}{\Delta t} = a$ the value of the acceleration characterises the intensity of the *electron process*, while the time system is quasi constant:

$$i \cong const.$$

With reference to Section 18, the *Quantum Membrane* is the drive of the *neutron collapse*. Ref S.18

The *blue shift* impact of the electron process is equal to the gained energy surplus of the *Quantum System of Reference,* the *Quantum Membrane*.

The impact of the electron process to the *Quantum Membrane* in absolute terms is:

$$\frac{dm_o c^2}{dt_i \cdot \varepsilon_x}\left(1-\sqrt{1-\frac{(c-i)^2}{c^2}}\right) = \frac{dn}{dt_i \cdot \varepsilon_x}q \qquad \text{24B3}$$

A certain value of mass change keeps balance with an equal number of quantum impact (in conventional terms: photons) in various intensity circumstances. The variety at the right hand side of the equation results in various frequency values.

dm_o – is the mass change in acceleration for $(c-i)$,

dt_i – denotes the time system of the electron process, the sphere symmetrical expanding acceleration at constant $i = \lim a\Delta t = c$ speed for infinite time; ε_x – is the intensity of the electron process of a certain material; n – is the number of the quantum impact.

$\varepsilon_x \ne (1/dt_i)$ the intensity depends on the characteristics of the element.

The standard electron impact is: $\dot{m}_i c^2 \frac{1}{\varepsilon_i}\left(1-\sqrt{1-\frac{(c-i)^2}{c^2}}\right)=f_i \frac{1}{\varepsilon_i} q$; and $dt_i = \frac{1}{\varepsilon_i}$

The electron process for a certain element, in its intensity form, is:

24B4
24B5

$$\dot{m}_i c^2 \frac{1}{\varepsilon_x}\left(1-\sqrt{1-\frac{(c-i)^2}{c^2}}\right)=f_i \frac{1}{\varepsilon_x} q \quad \text{or} \quad \dot{m}_x c^2\left(1-\sqrt{1-\frac{(c-i)^2}{c^2}}\right)=f_x \cdot q$$

where $\dot{m}_x = \frac{dm_o}{dt_i \varepsilon_x}$; and $f_x = \frac{dn}{dt_i \varepsilon_x}$

$\varepsilon_x = \varepsilon_i \cdot \varepsilon_e$ - we can call the electron intensity conditionally this way, where ε_i belongs to the $i = \lim a\Delta t = c$ time-frame; ε_e characterises the acceleration of the electron process of a certain x element.

With reference to a single energy quantum: the work, provided by the electron process (*blue shift*) corresponding to the energy of a single quantum – capable to impact the *Quantum System of Reference*, loading the *Quantum Membrane* is:

24C1

$$\mu c^2\left(1-\sqrt{1-\frac{(c-i)^2}{c^2}}\right)=q$$

24C2 $\frac{dm_i}{dt_i dn}=\mu$ standard electron mass change in balance with the energy of a single quantum impact ($dn=1$)

24C3

$$\dot{m}_{qe} = \mu\sqrt{1-\frac{(c-i)^2}{c^2}} \geq q$$ is the *quantum entropy*

24C4 *Quantum Membrane* can be loaded if: $\mu c^2 - \mu c^2\sqrt{1-\frac{(c-i)^2}{c^2}} > q$

The loaded *Quantum Membrane* means:

24C5

$$\frac{d(\mu \cdot n)c^2}{dt_i \cdot \varepsilon_x}\left(1-\sqrt{1-\frac{(c-i)^2}{c^2}}\right)=\frac{dn}{dt_i \cdot \varepsilon_x} q$$

The loaded *Quantum Membrane* initiates the collapse of the *quantum entropy.* It cannot withstand the quantum “pressure”, the drive of the re-transformation of the *quantum energy* into mass.

ε_x – the *intensity* of the electron process in 24C5 has its distinguishing importance: the infinite variety of the process provides infinite *variety* of load and the corresponding response of the *Quantum Membrane*.

<u>*Blue shift* impacts:</u>

- We take *electron process* with mass change values of $\dot{m}_{i1}$ and $\dot{m}_{i2}$ of the same element. The values of the acceleration for both masses of the same element are equal ($a_1 = a_2$ or $\varepsilon_1 = \varepsilon_2$).

  The mass relation however is: $\dfrac{\dot{m}_{i1}}{\dot{m}_{i2}} = x$ 24D1

  The relation of the frequencies will be:

  for mass change $\dot{m}_{i1}$: $$\frac{dm_1 c^2}{dt_i}\left(1-\sqrt{1-\frac{(a\Delta t)^2}{c^2}}\right) = \frac{dn_1}{dt_i}q = f_1 \cdot q \qquad \text{24D1}$$

  for mass change $\dot{m}_{i2}$ also: $$\frac{dm_2 c^2}{dt_i}\left(1-\sqrt{1-\frac{(a\Delta t)^2}{c^2}}\right) = \frac{x \cdot dn_1}{dt_i}q = x \cdot f_1 \cdot q \qquad \text{24D2}$$

  The frequencies are obviously *equal*, since the intensities of the mass change are equal just of broader scope. ($a_1 = a_2$ is equivalent to $\varepsilon_1 = \varepsilon_2$)

  The "fictive" durations of the two processes are different and proportional to the mass relations. The fictive duration also may mean the broadness of the same mass change impact in parallel: the width of the same frequency.

- We take the *electron process* of two different elements, but with equal mass change values of $\dot{m}_{iA} = \dot{m}_{iB}$. The acceleration and the intensity values of the mass change of different elements are different ($a_1 \neq a_2$ and $\varepsilon_1 \neq \varepsilon_2$):

  $$\frac{dm_A}{dt_i \varepsilon_A} \neq \frac{dm_B}{dt_i \varepsilon_B} \text{ and in other form: } \frac{\dot{m}_{iA}}{\varepsilon_A} \neq \frac{\dot{m}_{iB}}{\varepsilon_B} \text{ and } \frac{\varepsilon_A}{\varepsilon_B} = y \qquad \text{24E1}$$

  The *blue shift* and the generating frequencies of equal mass values of the electron process of two different elements are different:

  $$\frac{\dot{m}_A c^2}{dt_i \varepsilon_A}\left(1-\sqrt{1-\frac{(c-i)^2}{c^2}}\right) = \frac{dn_A}{dt_i \varepsilon_A}q = f_{iA} \cdot q; \quad \text{and} \qquad \text{24E2}$$

  $$\frac{\dot{m}_B c^2}{dt_i \varepsilon_B}\left(1-\sqrt{1-\frac{(c-i)^2}{c^2}}\right) = \frac{dn_B}{dt_i \varepsilon_B}q = f_{iB} \cdot q \qquad \text{24E3}$$

  The relation of the frequencies will be:

  $\dfrac{\dot{m}_{iA}}{\varepsilon_A} \neq \dfrac{\dot{m}_{iB}}{\varepsilon_b}$ and $f_{iA} \neq f_{iB}$ apparently, as per 24E1: $f_{iB} = y \cdot f_{iA}$

The time count of the *electron process* is constant ($dt_i = const$) for all elements, but the frequencies are *different*, because the intensities of the mass change vary.

During the <u>*neutron process*</u>, the mass in sphere symmetrical accelerating collapse, cannot withstand the increased (by the electron process) energy potential of the *Quantum Membrane*.

Ref S.18

Energy is (re-)transforming into mass. The driving force (of the *red shift*) of the neutron collapse is the impact from the *Quantum Membrane* (loaded by the *blue shift* of the electron process). The neutron process, the sphere symmetrical accelerating collapse of the *quantum entropy* happens under "external" effect.

24F1

$$w_n = \frac{dm_{qe}}{dt_i \varepsilon_n}\left(\frac{1}{\sqrt{1-\frac{(i-v)^2}{c^2}}}-1\right)$$

$(i-v)$ is the speed difference of the collapse to the *Quantum System of Reference*
- with the progress of the collapse $i \geq v \geq 0$ the speed difference is growing.

Ref 24A7

Similarly to 24A7, the mass and energy balance of the collapse is constant and equal at any stage of the process. The mass phase is growing and the energy phase is getting less and less.

$$\frac{\dot{m}_{qe}}{z_i} = \frac{\dot{m}_{qe-v}}{z_v} = \ldots = \frac{\dot{m}_{qe-y}}{z_y} \qquad \text{where} \qquad \dot{m}_{v-n} = \frac{dm_{qe}}{dt_{v-n}}$$

For comparative purposes, the event concentration and the time system for the *quantum entropy* is taken as $z_i = 1$ and $dt_i = 1$.

For the system with $(i-v)$ and $z_v = \sqrt{1-\frac{(i-v)^2}{c^2}}$

24F2

$$dt_v = \frac{dt_i}{\sqrt{1-\frac{(i-v)^2}{c^2}}} = \frac{dt_o}{\sqrt{1-\frac{(i-v)^2}{c^2}}\sqrt{1-\frac{i^2}{c^2}}}$$

24F3

and

$$\frac{\dot{m}_{qe-u}}{z_u} = \frac{dm_{qe}}{dt_u z_u} = \frac{dm_{qe}}{dt_i z_i \sqrt{1-\frac{(i-v)^2}{c^2}}}\sqrt{1-\frac{(i-v)^2}{c^2}} = \frac{dm_{qe}}{dt_i z_i} = \frac{\dot{m}_{qe}}{z_i}$$

There is no "escape" and no reflection of the impact from the surface of the collapse. All *red shifted* impact of the collision is incorporating into the collapsing mass. The incorporation increases the *event concentration* potential of the mass.

[We could use $(c-v)$ instead of $(i-v)$ as well. In this case, however, the end speed of the collapse would be $\lim v = 0$, while in the case of using $(i-v)$ it is conditionally taken as $v = 0$.]

With reference to 24F1

$$\frac{dmc^2}{dt_o\varepsilon_n}\sqrt{1-\frac{i^2}{c^2}}\sqrt{1-\frac{(c-i)^2}{c^2}}\left[\frac{1}{\sqrt{1-\frac{(i-v)^2}{c^2}}}-1\right]=\frac{dn}{dt_o\varepsilon_n}\sqrt{1-\frac{i^2}{c^2}}\cdot q \qquad \text{24F4}$$

With reference to the intensity of the proton process, ε_n characterises the acceleration of the collapse.

The balance between energy generation and its utilisation is:

Proton process energy generation

Neutron process energy utilisation at full collapse, when $v = 0$

$$\frac{dmc^2\left(1-\sqrt{1-\frac{i^2}{c^2}}\right)}{dt_o\varepsilon_p}=\frac{dn_p}{dt_o\varepsilon_p}q \qquad \text{24F5}$$

$$\frac{dmc^2\sqrt{1-\frac{(c-i)^2}{c^2}}\left(\sqrt{1-\frac{i^2}{c^2}}-1\right)}{dt_o\varepsilon_n}=\frac{dn_n}{dt_o\varepsilon_n}q \qquad \text{24F6}$$

From 24F5 and 24F6 it follows that the intensities of the proton and neutron processes may differ even in the most balanced circumstances.

The incorporation of the full absolute value of the energy during the *neutron collapse* is less than the energy generation during the proton process. At the end of the electron process, the *theoretical* mass status (since the measured value of the mass is its change) is equal to the *quantum entropy*:

$$\dot{m}_{qe}=\frac{dm}{dt_o}\sqrt{1-\frac{i^2}{c^2}}\sqrt{1-\frac{(c-i)^2}{c^2}}; \qquad \text{24G1}$$

Mass "status", at the end of the *neutron process* is:

$$\dot{m}_n=\frac{dm}{dt_o}\sqrt{1-\frac{(c-i)^2}{c^2}} \qquad \text{24G2}$$

Mass "status" at the end of cycle n is:

$$\dot{m}_n^n=\frac{dm}{dt_o}\left(1-\frac{(c-i)^2}{c^2}\right)^{\frac{n}{2}} \qquad \text{24G3}$$

S 24.2

24.1 External impact further loads the *Quantum Membrane*

The *Quantum Membrane* is capable of utilising the gathered energy. Any external load to the *Quantum Membrane* is an extra *blue shift* or electron process impact. The acting frequency is:

24H1 $$f_x = \frac{1}{q}\frac{dmc^2}{dt_i}\left(1-\sqrt{1-\frac{(c-i)^2}{c^2}}\right) = \frac{1}{q}\dot{m}_x c^2\left(1-\sqrt{1-\frac{(c-i)^2}{c^2}}\right); \text{ since } \frac{dn}{dt}q = f\cdot q$$

The higher is the intensity of the electron mass change (acceleration), the higher is the frequency, the number of the impacted photons.

If $\frac{c-i}{\Delta t_s} = a_s$ is the standard acceleration value, and $\frac{c-i}{\Delta t_x} = a_x$ is the impact; $x = \frac{\Delta t_x}{\Delta t_s}$

24H2 $$\frac{c-i}{\Delta t_x} = \frac{c-i}{x\Delta t_s} = \frac{1}{x}a_s \quad \text{and} \quad \varepsilon_x = \frac{1}{x}\varepsilon_s \text{ consequently:}$$

24H3 $$\frac{dmc^2}{dt_i\varepsilon_x}\left(1-\sqrt{1-\frac{(c-i)^2}{c^2}}\right) = \frac{dmc^2}{dt_i\varepsilon_s}x\left(1-\sqrt{1-\frac{(c-i)^2}{c^2}}\right) = \frac{dn}{dt_i\varepsilon_s}x\cdot q = f_x\cdot q$$

More frequency value means higher impacts for unit period of time – more "powerful" *Quantum Membrane*: $f_x = x\cdot f_s$ 24H4

S. 24.3

24.2 Motion with external energy

The *sphere symmetrical* expanding acceleration of mass m_o, up to speed *v*, needs the transformation of internal "mass energy" of:

24I1 $$\Delta E = \frac{dm_o c^2}{dt_o\varepsilon_o}\left(1-\sqrt{1-\frac{v^2}{c^2}}\right) = \frac{dm_o c^2}{dt_v\varepsilon_v}\left(1-\sqrt{1-\frac{v^2}{c^2}}\right)$$

In the case of *linear* acceleration of the same mass m_o up to the same speed *v*, but driven from external source, at constancy of the original mass ($m_o = const$), the external *red shift* impact need is:

24I2 $$\Delta w = \frac{dmc^2}{dt_o\sqrt{1-(v^2/c^2)}}\left(1-\sqrt{1-\frac{v^2}{c^2}}\right) = \frac{dn}{dt_o}q\left(1-\sqrt{1-\frac{v^2}{c^2}}\right)$$

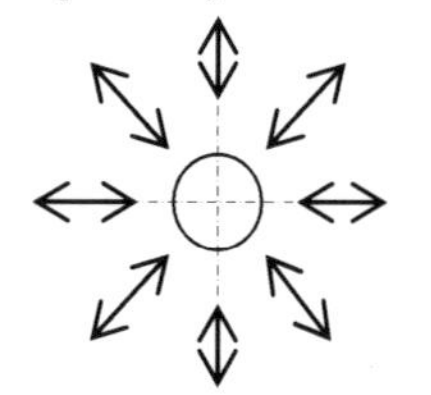

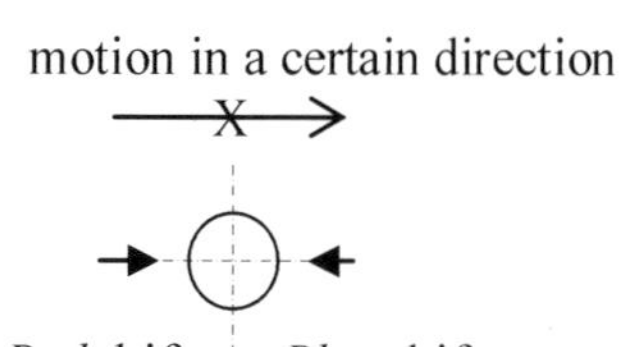

Fig.24.1

Fig 24.1

2412 is a classical *red shift*. The impacted *Quantum Membrane* gives off energy in the form of *red shift*. It can be written as

$$\frac{\dot{m}_o c^2}{\sqrt{1-\frac{v^2}{c^2}}} - \dot{m}_o c^2 = f_o \cdot q\left(1-\sqrt{1-\frac{v^2}{c^2}}\right) = \Delta f_o \cdot q \quad 2413$$

The right hand side of 2413 is a frequency difference, corresponding to the need of work of the acceleration of the mass, taken as constant.
The right hand side of 2413 is the *red shift*, the frequency change, measured within t_o, the system of reference of the mass acceleration.

For the demonstration of the conventional view, we have to slightly modify 2413. The right hand side of 2413 in fact is: $f \cdot H$. The dimension of *H*, the *Planck constant* is indeed *(kJoule.sec)*:

$$\frac{(d)m_o c^2}{\sqrt{1-\frac{v^2}{c^2}}} - (d)m_o c^2 = f \cdot q \cdot (d)t_o = f \cdot H \quad 2414$$

The missing time parameter within the conventional formula is "incorporated" into the *Planck-constant.*

Acceleration of mass in conventional terms ($m_o = const$) within the *Quantum System of Reference* loads the *Quantum Membrane*.

This load at conventional speed level of $v << c$ is different than at the level of *blue shift* generated by electron process. But acceleration of mass needs work anyway for balancing the *Quantum Membrane*, the *blue shift* generated by the motion.

24J1 The values of the *blue* and the *red shifts* indeed are: $f\left(1-\sqrt{1-\frac{v^2}{c^2}}\right)$

The work intensity for linear acceleration of constant mass is:

24J2
$$w_{motion} = w_{red} = w_{blue} = \pm f\left(1-\sqrt{1-\frac{v^2}{c^2}}\right)$$

the negative value means external energy need.

In the case of an already loaded *Quantum Membrane* by frequency f it is:

24J3
$$w_{motion} = w_{red} = w_{blue} = f \pm f\left(1-\sqrt{1-\frac{v^2}{c^2}}\right)$$

The frequency impact can be expressed with a certain mass balance of m_{xo} as well. In this regard, the work and the energy intensity relation is: $\dot{m}_o c^2 = f_o \cdot q$ (24J4)

24K1 From 24I3:
$$\dot{m}_o c^2\left(1-\sqrt{1-\frac{v^2}{c^2}}\right) = \dot{m}_{xo} c^2 \underline{\sqrt{1-\frac{v^2}{c^2}}}\left(1-\sqrt{1-\frac{v^2}{c^2}}\right)$$

this component is equivalent
to the remaining mass status of the transformation mass m_{xo} to speed v

24K2
$$m_{xo}c^2 = m_{xo}c^2\left(1-\sqrt{1-\frac{v^2}{c^2}}\right) + m_{xo}c^2\sqrt{1-\frac{v^2}{c^2}}$$

The relation of masses, one is accelerated with external energy, the other is the virtual accelerating source:

24K3
$$\underbrace{\frac{\dot{m}_o c^2}{\sqrt{1-(v^2/c^2)}} - \dot{m}_o c^2}_{\text{energy, transformed into mass virtual increase of mass}} = \underbrace{\dot{m}_{xo}c^2 - \dot{m}_{xo}c^2\sqrt{1-\frac{v^2}{c^2}}}_{\text{mass, transformed into energy, the source of the acceleration}}$$

from 24K1 follows: $\dot{m}_o = \dot{m}_{xo}\sqrt{1-\frac{v^2}{c^2}}$

And the continuity of the matter proves itself again:

24K4
$$\frac{m_o}{\sqrt{1-(v^2/c^2)}} + m_{xo}\sqrt{1-\frac{v^2}{c^2}} = m_o + m_{xo}$$

The sum of the virtual increase of the mass and the remaining mass of the acceleration is equal to the original values.

25

S
25

Quarks – transformation again

The equivalence of mass and energy relates to *process*: *transformation* of mass into energy, *re-transformation* of energy into mass, *impact* in form of *blue* and *red shifts* to and from the *Quantum Membrane*. Mass energy of $\dot{m}c^2$ equivalent to $n \cdot q$ means in fact the balance of a *process*, transformation and impact for a certain period of time.

The sphere symmetrical expanding acceleration and collapse have a specific feature: there is no "centre (at rest)" in these processes. The reason is very simple: Time cannot be defined without event. The question, therefore, is, what is inside the expanding or collapsing mass of matter?

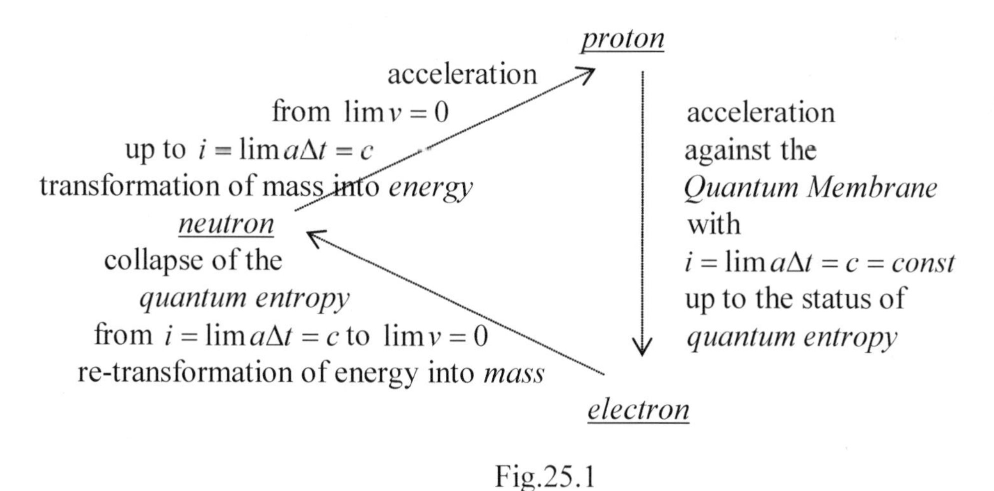

Fig.25.1

Fig.
25.1

The end statuses of the proton, electron and neutron processes are well defined. At the end of the electron process we have the status of the expanding mass with internal working capability incapable of withstanding the impact of the *Quantum Membrane*.

The electron process, the sphere symmetrical expanding acceleration for infinite time, the motion with $i = \lim a\Delta t = c$, is balanced by the *Quantum Membrane*. What happens once the electron process fully expires its balancing accelerating mass (-energy reserve) against the *Quantum Membrane*?

Mass will be collapsing and the neutron process – consolidation of the mass-energy balance under the effect of the *Quantum Membrane* – starts from the status of *quantum entropy*. The system of reference of the *quantum entropy* has $\lim \varepsilon_{qe} = 0$ – infinite low intensity and $\lim dt_{qe} = \infty$ – infinite long time system of reference. The re-transformation of energy into mass means also the change of the time system – the consequence of the motion!

With the decrease of the speed of the sphere symmetrical expanding acceleration from $i = \lim a\Delta t = c$ to less and less $v < i$ and $v << i$ values, the intensity of the process increases and the time system of the process slows down: the duration of the processes are getting shorter and shorter. The balance of the change in absolute mass values however is constant:

25A1 $$\frac{dm_{qe}c^2}{dt_{qe}\varepsilon_{qe}} = \frac{dm_{qe}c^2}{dt_v\varepsilon_v} = \frac{dm_i c^2}{dt_i\varepsilon_i} = \frac{dm_i c^2}{dt_v\varepsilon_v};$$

25A2 the balance of the relativistic "measured" masses is $\frac{\dot{m}_i c^2}{\varepsilon_i} = \frac{\dot{m}_v c^2}{\varepsilon_v}$.

Since: $i - v = \Delta v = a\Delta t$, the time and intensity relations are:

25A3
25A4 $$dt_i = \frac{dt_v}{\sqrt{1-(\Delta v^2/c^2)}} \quad \text{and} \quad \varepsilon_v = \frac{\varepsilon_i}{\sqrt{1-(\Delta v^2/c^2)}}$$

The intensity of the collapsing mass of the neutron is growing:

25A5 $$\frac{\dot{m}_v}{\varepsilon_v} = \frac{\dot{m}_i}{\varepsilon_v\sqrt{1-(\Delta v^2/c^2)}} \quad \text{and} \quad \dot{m}_v > \dot{m}_i$$

25A1 – 25A5 describe the re-transformation of quantum energy into mass. The value of the *"matter"* in total is the same, its mass and energy components, however, are changing. The sphere symmetrical accelerating collapse can stop, when the working capability of the mass within the atomic structure, can withstand the *red shift* impact of the *Quantum Membrane*. The *red shift* impact is generated by the *blue shift* of the electron process. The velocity of the collapse is constantly decreasing. The end status of the collapse is at $\lim u = 0$.

At $\lim u = 0$, the time flow is $\lim dt = 0$ (the shortest) and the intensity is $\lim \varepsilon_u = \infty$ (the highest). Just for good measure, the *intensity* of the neutron process at the start of the collapse – at the status of the *quantum entropy* – is the lowest and the duration of the event is the longest.

The intensity of the neutron process is permanently growing and the duration is permanently shortening. These changes cannot stop the collapse. The speed difference of the motion between the *Quantum System of Reference* and the front surface of the collapsing neutron is also growing.

The *red shift* impact from the *Quantum Membrane* permanently grows:

$$\dot{w}_{red-shift} = f \cdot q\left(1 - \sqrt{1 - \frac{(c-u)^2}{c^2}}\right) \qquad \text{25B1}$$

$$\dot{w}_{red-shift-u1} = f \cdot q\left(1 - \sqrt{1 - \frac{(c-u_1)^2}{c^2}}\right) \qquad \dot{w}_{red-shift-u2} = f \cdot q\left(1 - \sqrt{1 - \frac{(c-u_2)^2}{c^2}}\right)$$

at speed values $u_2 < u_1$ $\qquad \dot{w}_{red-shift-u2} > \dot{w}_{red-shift-u1}$

At the status of the collapse with $\lim u = 0$ the *red shift* impact will not stop. It will be acting even when the speed will not be decreasing any more.

There is no way for the collapse to continue: the speed is $\lim u = 0$. This is the natural conflict between the *red shift* impact from the *Quantum Membrane* and the growing internal energy (working capability) of the mass. This is the point when the internal energy of the mass prevails and the proton process, with the acceleration of the mass, starts.

With the increase of the sphere symmetrical expanding acceleration, contrary to the neutron process, the mass of the proton will be less and less, providing energy (with the transformation of mass into energy) to the *Quantum System of Reference*.

Following the methodology of the continuity of the matter,

the balance of absolute mass values for the proton process is: $\dfrac{dm_o c^2}{dt_o \varepsilon_o} = \dfrac{dm_o c^2}{dt_v \varepsilon_v}$ $\qquad$ 25C1

the balance of the measured relative masses is: $\dfrac{\dot{m}_o c^2}{\varepsilon_o} = \dfrac{\dot{m}_v c^2}{\varepsilon_v}$ $\qquad$ 25C2

Since: $v - \lim 0 = \Delta v = a\Delta t$, the time and intensity relations are:

25C3 $$dt_v = \frac{dt_o}{\sqrt{1-(\Delta v^2/c^2)}} \quad \text{and} \quad \varepsilon_v = \varepsilon_o\sqrt{1-\frac{\Delta v^2}{c^2}}$$

The intensity of the expanding acceleration of the mass of the proton is decreasing (transformation of mass into energy):

25C4 $$\frac{\dot{m}_v}{\varepsilon_o\sqrt{1-(\Delta v^2/c^2)}} = \frac{\dot{m}_o}{\varepsilon_o} \quad \text{and} \quad \dot{m}_v < \dot{m}_o$$

The end status of the proton process is the sphere symmetrical expanding acceleration with speed $i = \lim a\Delta t = c$. And the electron process starts, which is *blue shift* impact to the *Quantum Membrane*, the drive of the neutron collapse.

Will the *proton process* with mass less and less mean the collapse of its size? Will the *neutron collapse* with growing mass mean size growth?

Our conventional view dictates that getting less in mass will result in less size and growing in mass more shall result is increased size. In this way the sphere symmetrical expanding acceleration of mass would mean the sphere symmetrical collapse in size and the sphere symmetrical collapse of mass the sphere symmetrical expansion in size.

In the case of the sphere symmetrical expanding acceleration and collapse, the main point is the transformation. Space coordinates are measurements, function of time flow. *Zero* time flow would mean no event and *zero* extension. The event concentration at $\lim u = 0$ is infinitely high with $\lim \Delta t = 0$ (infinitely "short" time measurement) and $\lim \Delta l = 0$ (infinitely "short" distance) measurements. $i = \lim a\Delta t = c$ means infinite long time flow and infinite large size measurement.

There is no significance in what the size of the mass is, or what the direction of the change of the space measurement is within the system of reference of the mass in transformation.

Does the smallest mass, shortest time, highest event concentration exist? Is there any component or particle within the material world that we could call the last one?

This cannot be the case!

From the philosophical point of view, this would automatically mean that *world* has its end and that infinity as such only exists in mathematical formulas. The proof, however, comes from the facts of physics: The mass change of the *quantum entropy* status is

$$\lim \frac{dmc^2}{dt_i\varepsilon_q}\left(1-\sqrt{1-\frac{(c-i)^2}{c^2}}\right)=0 \qquad \text{25D1}$$

where time and intensity parameters relate to $\lim i = c$, and the acceleration itself is of infinite small value of $\lim(c-i)=0$.

$$\lim dt_i = \lim \frac{dt_o}{\sqrt{1-(i^2/c^2)}} = \infty; \quad \text{and} \quad \lim \varepsilon_i = \lim \varepsilon_o \sqrt{1-(i^2/c^2)} = 0 \qquad \text{25D2, 25D3}$$

None of the above parameters is of finite value. And one parameter is still missing: the variety of the mass change. (Time and intensity values relate to imaginary rest status, which is theoretically correct, but "no event" at absolute rest status would mean "no time".) The mass in 25D1 means de facto a change. In this case the diapason of the mass change is infinite:

$$\text{from } \lim \dot{m} = 0 \text{ up to } \lim \dot{m} = \infty \qquad \text{25D4}$$

This circumstance can be characterised with event concentration z_q of the *quantum entropy*, and the value of the intensity of the process in 25D1 varies from $\lim \varepsilon_q = \varepsilon_i z_q = 0$ up to $\lim \varepsilon_q = \varepsilon_i z_q = \infty$: $\quad 0 = \lim \varepsilon_q = \infty \qquad$ 25D5

Photons have *zero* mass indeed but their equivalent energy quantum is real and different than *zero*:

$$\frac{dmc^2}{dt_v\varepsilon_v}\left(1-\sqrt{1-\frac{v^2}{c^2}}\right)=\frac{dn}{dt_v\varepsilon_v}q \qquad \text{25D6}$$

The uniqueness of the transformation of mass into energy is that the values of *n* in 25D6, the number of the generated photons, is always an *integer*.

25.1 Quarks

S. 25.1

The sphere symmetrical expanding acceleration and collapse are only understandable if the internals of *particles* (proton, neutron, electron and others) are either in *plasma* status, with reference to Section 20 – in infinite *blue shift* conflict and motion, or the universal transformation of mass into energy and the re-transformation of energy into mass continues "without end" within the particles, which are in fact processes themselves.

Ref S.20

Quarks, identified by Murray Gell-Mann and George Zweig in 1964, as particles – are the measured effects (of mass change) of the mass-energy transformation within protons and neutrons. The energy, equivalent to the energy of "*quantum*" was called "*gluon*" within the energy field of hadrons. The *Quantum System of Reference* is the same within and outside the *hadrons*, but for the adequacy of the energy balance within the proton and neutron particles, we can accept that the *Quantum System of Reference* within the hadrons is called *Gluon System of Reference* and the *Quantum Membrane* – the *Gluon Membrane*.

There are 6 measured types (flavours) of *quarks*, with measured minimal, standard and maximal mass effects presented in Table 25.1. (Mass cannot be measured rather the consequence of the mass change.) Therefore, it is more appropriate to call these "mass values" the *measured effects* of the mass-energy transformation process within the hadrons.

Down	**Strange**	**Bottom**	**Top**	**Charm**	**Up**
measured mass (impact) $\dot{m}$ [MeV]					
3.4	70	4,130	169,100	1,160	1.4
4.8	**104**	**4,200**	**171,200**	**1,270**	**2.4**
6.6	130	4,370	172,400	1,340	3.3

Table 25.1

Table 25.1

The neutron, the proton and the electron processes are in mass and energy balance, internally and externally.

- externally – means within the atomic structure;
- internally – means within the neutron, the proton and the electron particles themselves.

Neutrons and *protons* are the transformation of mass and energy, representing a certain balance between sphere symmetrical expanding acceleration – measured as *Top-Charm-Up* (TCU) "quarks" – and sphere symmetrical accelerating collapse – measured as *Down-Strange-Bottom* (DSB) "quarks". *Electrons* are the "drive" between the neutron and proton processes. The electron process does impact the *Quantum Membrane.*

In the *proton process* the dominant is the expansion, the transformation of mass into energy.
In the *neutron process*, the dominant is the collapse, the re-transformation of energy into mass.

The neutron process is driven by external source, the *red shift* of the *Quantum Membrane*, generated by the *blue shift* of *electrons*.

We can call these statuses measured quarks U, C, T and D, S and B, but obviously we measure the effect of the mass change.

The TCU chain “of quarks” is dominant with sphere symmetrical expanding acceleration and gives off energy. The DSB chain “of quarks” is dominant with sphere symmetrical accelerating collapse and with incorporation of energy. Without external impact the collapse does not work, since the *Gluon System of Reference*, the system of reference of equal energy *gluons*, without impact does not have the *membrane function*. For establishing the *Gluon Membrane* within the neutron, external impact is necessary. Otherwise there is no energy-mass-energy communication between the TCU and the DSB chains. The *Gluon System of Reference*, the result of the TCU chain, would exist, but, without impact, the *Gluon Membrane* would not work.

The internal neutron process works only under the *red shift* impact of the *Quantum Membrane* generated by the *blue shift* of the electron process. The collapse, the accumulation of the working capability in the form of mass, ends with a kind of saturated status. The *Bottom quark*, full of mass “energy”, is ready for sphere symmetrical expanding acceleration, the measured *Top quark*. The acceleration starts with *Top* and ends with *Up*. The *Up* status, the loss of all available energy in fact leads to the measured *Down quark* and the cycle with the collapse starts again.

The collapse of the DSB chain goes through measured *Down, Strange* and *Bottom* quarks. The sequential TCU chain continues the acceleration with *Top* quark via *Charm* and ending at *Up* status. The next status after *Up* again would be *Down* and the process would be repeating in an infinite number of cycles.

Quark *types* represent different statuses of the sphere symmetrical expanding acceleration and accelerating collapse processes. The DSB chain results in TCU chain and the acceleration, the transformation of mass into energy, starts again. As Section 18.2 proves, the energy part of the process grows step by step and the re-transformation into mass goes with mass deficit. This difference – for the benefit of the energy part of the process – is the drive in accordance with the *Second Law of Thermodynamics*.

Ref. S.18.2

The DSB and the TCU chains can exist only together. The dominant DSB in the neutron process becomes dominant TCU in the sequential proton process. The expanding acceleration process will be more dominant in the proton process than the collapse. The result is surplus in *gluons* – energy quantum. This energy quantum "leaves" the balanced proton process and joins the *Quantum System of Reference.*

S. 25.1.1 *25.1.1.* ***Protons, neutrons and electrons are processes***

The accumulation in the neutron process mass-"energy" initiates the proton process: the transformation of mass into energy. The electron process is the drive of the neutron process, the re-transformation of energy into mass. The process goes in infinite cycles, resulting different elements and as *events*, establishing time.

Ref S.17 S.18

With reference to Section 17 and 18, the descriptions of the *elementary processes* are:

25E1 The *proton process* of cycle n is: $\dot{m}c^2\left(1-\sqrt{1-\frac{i^2}{c^2}}\right)$

25E2 The *electron process* of cycle n is: $\dot{m}c^2\sqrt{1-\frac{i^2}{c^2}}\left(1-\sqrt{1-\frac{(c-i)^2}{c^2}}\right)$

The *neutron process* of cycle n is:

25E3
$$\frac{\dot{m}c^2\sqrt{1-\frac{i^2}{c^2}}\sqrt{1-\frac{(c-i)^2}{c^2}}}{\sqrt{1-\frac{(c-u)^2}{c^2}}}-\dot{m}c^2\sqrt{1-\frac{i^2}{c^2}}\sqrt{1-\frac{(c-i)^2}{c^2}}=\dot{m}c^2\sqrt{1-\frac{(c-i)^2}{c^2}}\left(\sqrt{1-\frac{i^2}{c^2}}-1\right)$$

25E1, 2 and 3 show that the new $n+1$ cycle starts with less proton mass:

25E4 The measured proton mass of the start of cycle $n+1$ (equal to the measured mass value of the neutron status at cycle n) is: $\dot{m}\sqrt{1-\frac{(c-i)^2}{c^2}}$

25E5 The measured proton mass of cycle n was: $\dot{m}$

The electron process loads the *Quantum Membrane* and initiates the neutron collapse. With reference to 24B5, the load, relative to a single quantum, corresponds to

Ref 24B5

$$\mu\left(1-\sqrt{1-\frac{(c-i)^2}{c^2}}\right)\cdot c^2=q$$

and, with reference to 24C4, the load is only possible if the energy of the electron impact is: Ref 24C4

$$\mu c^2 - \mu c^2 \sqrt{1 - \frac{(c-i)^2}{c^2}} \geq q$$

Once the change of the electron mass achieves the value of the *quantum entropy*, it will be incapable of loading the *Quantum Membrane*:

$$\dot{m}_{qe} = \mu \sqrt{1 - \frac{(c-i)^2}{c^2}}$$ is the *quantum entropy*

There is quasi "balance – unbalance" within the elementary processes, which is the drive of *events* in *time. Quarks* in protons, neutrons and electrons are *events*, rather than static particles, representing the endless *...mass-energy-mass-energy-mass...* process of change.

The change is in quasi balance: the mass, in transformation into energy is equal to the energy re-transforming into mass and vice versa. The key point is the general rule of *entropy*: the change can never reach *zero*. The same energy or the same mass cannot be in two statuses at the same time. Either mass or either energy, but never both. And the *quantum entropy* as such becomes the *engine* of events and with that *time*. The *matter* permanently compensates for the unbalance and the *world* moves ahead in time. (*Matter* also compensates for external impacts: electron process, loaded by "man-made" impacts, works the same way – with the required benefit.)

The loaded *Quantum Membrane* with reference to 24C5 means: the load depends on n, the value and ε_x, the intensity of the impact

$$\frac{d(\mu \cdot n)c^2}{dt_i \cdot \varepsilon_x}\left(1 - \sqrt{1 - \frac{(c-i)^2}{c^2}}\right) = \frac{dn}{dt_i \cdot \varepsilon_x} q$$

Ref 24C5

Proton and neutron processes, with electron process drive (and all *quark* chains in elementary particles) compose one and the same unity of the matter.

$$\frac{dm}{dt_p \varepsilon_p}\left(1 - \sqrt{1 - \frac{v^2}{c^2}}\right) = \frac{dm}{dt_n \varepsilon_n}\left(\frac{\sqrt{1 - \frac{i^2}{c^2}}\sqrt{1 - \frac{(c-i)^2}{c^2}}}{\sqrt{1 - \frac{(c-v)^2}{c^2}}} - \sqrt{1 - \frac{i^2}{c^2}}\sqrt{1 - \frac{(c-i)^2}{c^2}}\right) \quad \text{25F1}$$

but $(c - v) = i$ and:

$$\frac{dm}{dt_p \varepsilon_p} = \frac{dm}{dt_n \varepsilon_n}\sqrt{1 - \frac{(c-i)^2}{c^2}}; \quad 1 = \frac{dt_p \varepsilon_p}{dt_n \varepsilon_n}\sqrt{1 - \frac{(c-i)^2}{c^2}}; \quad \varepsilon_e = \frac{\varepsilon_p}{\varepsilon_n}\sqrt{1 - \frac{(c-i)^2}{c^2}};$$

25F2
25F3
25F4

25F5 From 25F3 and 25F4: $\varepsilon_e \dfrac{\varepsilon_p}{\varepsilon_n} = \dfrac{dt_p \varepsilon_p}{dt_n \varepsilon_n}$ $\varepsilon_e \dfrac{\varepsilon_n}{\varepsilon_p} = 1$ <u>*the same event!*</u>

The mass change chain is:

25G1

$$m \rightarrow \underbrace{m\sqrt{1-\frac{i^2}{c^2}}}_{\text{proton}} \rightarrow m\sqrt{1-\frac{i^2}{c^2}}\sqrt{1-\frac{(c-i)^2}{c^2}} \rightarrow m\sqrt{1-\frac{(c-i)^2}{c^2}} \rightarrow$$

proton — electron — neutron — proton

$$\leftarrow m\sqrt{1-\frac{i^2}{c^2}}\sqrt{1-\frac{(c-i)^2}{c^2}}\sqrt{1-\frac{(c-i)^2}{c^2}} \leftarrow m\sqrt{1-\frac{i^2}{c^2}}\sqrt{1-\frac{(c-i)^2}{c^2}} \leftarrow$$

electron

neutron

$$\rightarrow m\sqrt{1-\frac{(c-i)^2}{c^2}}\sqrt{1-\frac{(c-i)^2}{c^2}} \rightarrow m\sqrt{1-\frac{i^2}{c^2}}\sqrt{1-\frac{(c-i)^2}{c^2}}\sqrt{1-\frac{(c-i)^2}{c^2}}$$

proton

Fig 25.2

Fig. 25.2

Fig.25.2 demonstrates that the *neutron process* is identical to the sequential *proton process* of the next cycle – just the mass change is in the opposite direction.

The electron process keeps balance with the *Quantum Membra*ne, until

25H1
$$\frac{dmc^2}{dt_i \cdot \varepsilon_i}\left(1-\sqrt{1-\frac{(c-i)^2}{c^2}}\right) \geq \frac{dn}{dt_i \cdot \varepsilon_i} q$$

25H2 If the result is $\dfrac{dn}{dt} = 1$, the process cannot impact the *Quantum Membrane*.

25H3 From 25H2 follows that $\dfrac{dmc^2}{dt_i \varepsilon_i}\left(1-\sqrt{1-\dfrac{(c-i)^2}{c^2}}\right) = \dfrac{1}{dt_i \varepsilon_i} q$

25H4 For simplicity's sake: $dt_i = 1$; $\varepsilon_i = 1$, and

25H5
$$\frac{dm}{dt_i} = \dot{m}_i$$

25H5 is the conventional meaning of the mass value, but in fact it characterises the *change* of the mass within the system of reference of the measurement. Mass cannot be measured – just the effect of the mass change. The consequence of this fact is that we cannot establish the real mass of the particles. We can, however, measure similar effect of different mass values: $$\dot{m}_i c^2\left(1-\sqrt{1-\frac{(c-i)^2}{c^2}}\right)=q_e$$ 25H6

Quantum entropy does not correspond to an exact mass value. The effect of the mass change keeps balance with the *Quantum Membrane*. (Meaning: the mass can vary.)

25.2 *Quark*-process

S. 25.2

The objective of the description below is not speculation rather the demonstration of a certain concept.
It can be stated without doubt: *Quarks* cannot be static particles. They are part of the matter, which exists in motion and time. *Quarks* shall be vital parts of the sphere symmetrical expanding acceleration and accelerating collapse processes.

The force, which keeps particles together, is the *continuity law of matter* within each single proton, neutron, electron and others. (Particles are events in space and time.) There is no "empty space" as such or vacuum within the matter. The matter exists either in mass or energy. The collapse in principle creates "emptiness" and the acceleration, refills this "empty space" with "energy".

The *continuity law of matter* is based on *Strong* and *Weak Forces, strong* and *weak interactions* between energy and mass. *Strong interaction* represents transformation. *Blue* and *red shift* impacts belong to *Weak interactions.*

The proton is energy provider. The neutron is an energy "user", the transformation of energy into mass through collapse. The TCU chain in the proton is weaker and weaker and the DSB chain within the neutron process is stronger and stronger. The TCU chain provides more energy through acceleration (transformation) than the DSB chain can utilise with collapse. This energy benefit is the "feeding energy source" of the *Quantum (or Gluon) System of Reference.*

The significance of the *Down-Strange-Bottom-Top-Charm-Up-Down-Strange-Bottom...and so on* process is, that the measured *Up* and *Down* statuses, both refer to infinite long time systems.

The *Up* status is the end of the acceleration and the *Down* status is the start of the collapse, both at speed level of $i = \lim c$. The most important indicator of these two statuses is *time*, the infinitely long time period of their appearance (relative to other more intensive other processes, with shorter existence).

The measurability is the key point in detection of quarks. At speed $i = \lim c$ of the acceleration the time frame is infinitely long and the intensity of the process is infinitely weak ($\lim \varepsilon = 0$) – therefore, the measurement is difficult. The *last* and *first* (or *first* and *last*) measurable mass values are the *Up* and *Down* (or *Down* and *Up*) processes (statuses). The best measurement conditions are obviously for *Top* and *Bottom* processes (statuses), but these are available for less (limited) time.

S. 25.2.1

25.2.1 The relation of Strong and Weak Forces

The *electron* process is sphere symmetrical expanding acceleration for infinite time. The measured statuses are D_D_D. This is understandable since the sphere symmetrical expanding acceleration is of low intensity.

The *neutron* process is of two cycles:
- a closed one with D-S-B *collapse* and with T-C-U *expansion.*
  [There should be two processes here, since the D-S-B is with measured (+) sign, but the T-C-U is with measured (–) sign.]
  They are of different direction. This also means the neutron process is not just of full collapse – just as the proton process is not of full expansion.)
- an open cycle with *collapse* D-S-B, to which the neutron process does not provide the expansion part. (That is why the neutron process is with collapse dominance.)

The missing expansion demand of the open collapse within the neutron process is covered by the not used expansion within the *proton* process. The *proton* process is also composed of two parts:
- an open cycle with T-C-U *expansion,* responding to the not addressed collapse within the neutron:
- a closed one with T-C-U *expansion* and D-S-B *collapse.*

The sphere symmetrical expanding acceleration of the open cycle of the proton and the uncovered collapse of the neutron processes are bound together as *Strong Force.* The processes within the closed internal cycles are bound together the same way, by *Strong Interrelation.* This force keeps the measured quarks within the *hadrons* inseparable from each other.

Strong Interrelation keeps not just the *hadrons* together, but makes the proton-neutron relation effective and inseparable in any circumstances. Proton and neutron cannot exist without each other.

The proton process is a clear loss on mass change.
- the open cycle obviously transforms the mass into energy;
- the closed cycle is also getting less (following the entropy rule).

With the loss of the mass change, the measurement of the proton process will be dominant with quarks U_U_D. (The measurement of the expansion gives U and the collapse gives D. This is no other than the effect of the slow down of the process, while the speed of the expanding acceleration reaches $i = \lim a\Delta t = c$. Proton expires all its mass "energy capability".)

The *electron process* is also of collapse and expansion with open and closed cycles. The process is of low intensity, giving off all possible mass energy, reaching the status of *quantum entropy*.

The measurements will give D_D_D. The open cycle results in *Weak Interaction (Force)* – the *blue shift*. The mass change effect is very weak and of constant time frame.
(The electron process is expansion, like the proton process, just of constant timeframe.)

The electron process drives the collapse in both, the proton and the neutron processes, just the dominance in the proton is the expansion and in the neutron the collapse!

As end of the *electron* process, the neutron process starts. The neutron process, with its open and closed cycles, has two external impacts:
- the coverage of the T-C-U expansion to the open D-S-B from the proton, and
- the utilisation of the *blue shift* impact of the electron.

The impacting electron process represents a change from (–) to (+). The measurement of D of the open cycle of the electron turns to U, since it produces *blue shift*.

While the mass change of the closed cycle of the neutron process is getting less and less, the mass change within its open cycle is growing. This is the reason why the changing mass at the end stage of the neutron process will be less than the original proton mass change was.

The neutron process is the increase in mass – collapse, the incorporation of the T-C-U energy of the proton open cycle. The *neutron* will be re-born as *proton* and with its sphere symmetrical expanding acceleration the elementary cycle continues without end.

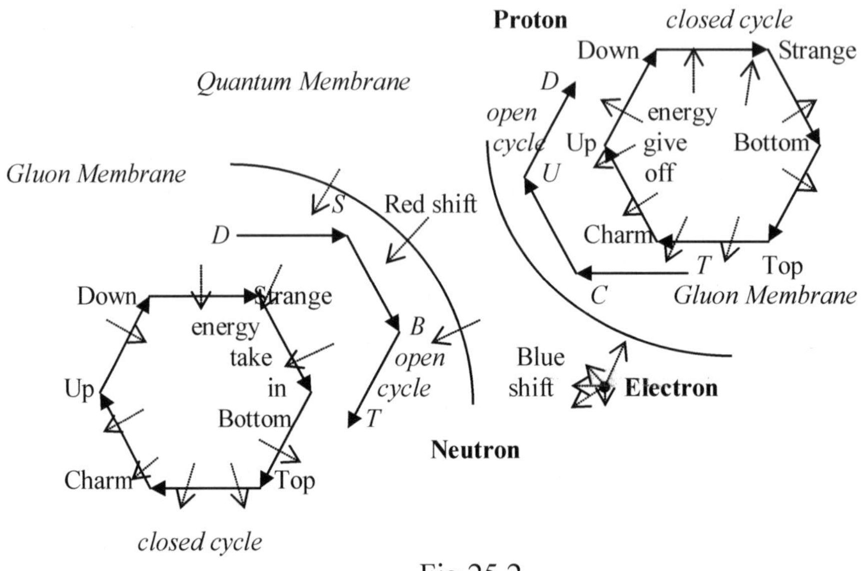

Fig. 25.2

Fig.25.2

The DSB chain is the dominant in the neutron process and the TCU chain is the dominant in the proton process. This is the main feature and reason why the proton is proton and the neutron is neutron. *Quarks* do not exist out of the proton and neutron processes. They are establishing the proton and neutron processes themselves.

Table 25.2 compares the availability of quarks to be detected relative to their motion and time flow. Electrons can be measured for long. Their time flow is $\lim t = \propto$. U and D quarks are approaching and leaving behind $\lim t = \propto$ in their process. The last row in Table 25.2 refers to the time frame of the possible detection.

Electron	*Neutron*			*Proton*			*Electron*	*Neutron*
< U	**U**	**C**	**T**	**B**	**S**	**D**	**< U**	**U**
< U	**D**	**S**	**B**	**T**	**C**	**U**	**< U**	**D**
< U	**D**	**S**	**B**	**T**	**C**	**U**	**< U**	**D**
long	long	limited	short	short	limited	long	long	long

Table 25.2

Table 25.2

The length of the arrows in Fig.25.3 demonstrates the dominance of the process within the particles and includes the effect of the open cycles. The electron process is sphere symmetrical expanding acceleration against the *Quantum Membrane*, motion with $i = \lim a\Delta t = c$.

The presentation of all *quarks* within the electron is a kind of presentation of the process of the changing mass. At the last stage of the proton process, all *quarks* will be close to the same, limited mass.

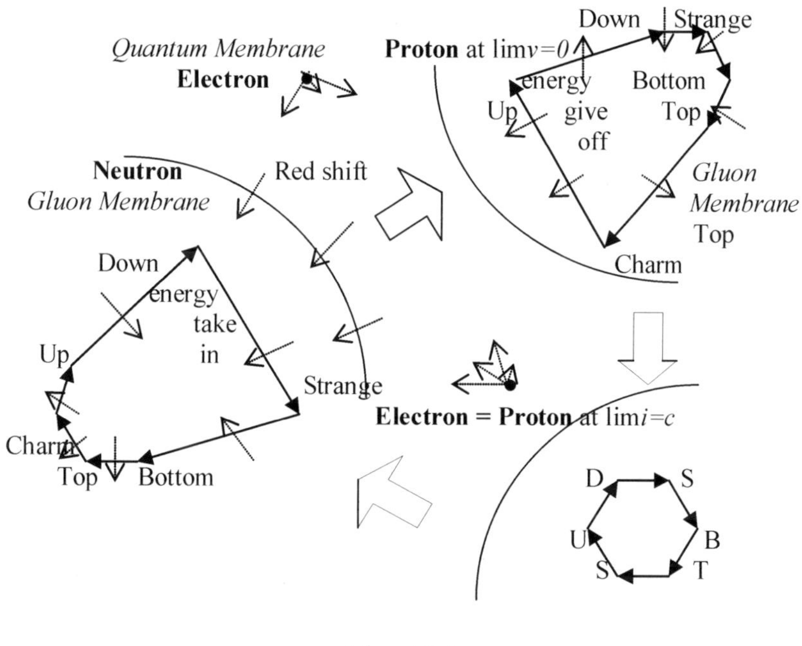

Fig. 25.3

Fig 25.3

The fact that *particles* and *antiparticles* have been found proves that the detection of *quarks* is much wider than the classification given. There are detections with mass effect, equivalent to the *Up*, *Charm* and *Top* statuses within the collapse dominance neutron process. In the same way, there are detections, corresponding to *Down*, *Strange* and *Bottom* quarks within the expansion dominance proton process.

The expansion and collapse is a continuous process, going through all mass process statuses and effects. The detection of similar effects – *quarks* and *anti-quarks* – within the proton and neutron processes is thus understandable. Therefore, with reference to Table 25.1 with mass (effect) values, it is better to present the statuses via a balanced circle. There are many more mass effects in this circle, which have not been detected.

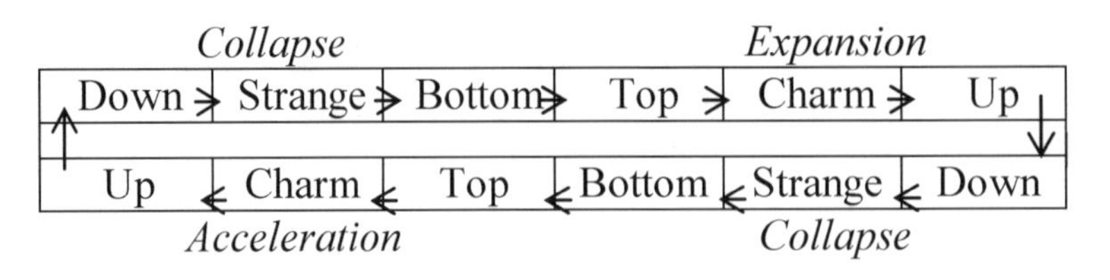

Table 25.3

Table 25.3

The quarks in Fig.25.3 go through all statuses in collapse and acceleration, just the dominance is different in the proton and neutron processes.

S. 25.2.2

25.2.2. Quark process intensities

Proton process starts with infinite ε_p intensity. Proton accumulates all mass energy of the neutron process. Infinite intensity relates to infinite short duration.
The end state of the *neutron process* is similar: infinite high intensity and infinite short duration. The accumulated mass energy prevails and the sphere symmetrical expanding acceleration – as proton process – starts: previous neutron continues as proton.

The first (in the proton process) and the last (in the neutron process) measurable *quarks* of these events are *Top* and *Bottom* – rarely measured, short lived processes. *Up* and *Down* processes belong to high speed accelerations with low intensity and close to infinite durations, therefore, with measurable characteristics for long. *Charm* and *Strange* are between the two ends.

The internal neutron process works only under the *red shift* impact of the *Quantum Membrane* generated by the *blue shift* of the electron process.

The collapse, the accumulation of the working capability in the form of mass, ends with a kind of saturated status. And the expanding acceleration starts as measured *Top* and ends as measured *Up*.

Quark *types* represent different "statuses" of the acceleration and collapse processes. The *Second Law of Thermodynamics* does not allow processes to happen without increase of *entropy*. Since the entropy grows with each process, this chain of mass change goes only with energy impact – generated by the chain itself = as electron process!

Quarks in the proton process lose "mass" energy to the benefit of quarks within the neutron process. The *Quantum Membrane* (result of the electron process) impacts the *Gluon System of Reference* of neutrons and the collapse continues. The *red shift* impulse of the neutron collapse originates from the proton process as electron process *blue shift* impact.

Electron is measured with 3 *Up* quark processes. The proton process is measured with 3 quarks: *Up-Up-Down*. It cannot be differently, since any measured "mass" loss behind the *U_U_D* status automatically gives electron status. Stabilised neutron process is mainly measured with 3 quark processes: *Up-Down-Down*. *U_U_D* also can characterise the neutron process, but in conventional physics they call it as *anti*-proton. *Anti*-neutrons (composition of *UDD*) can also be measured in the proton process.

There is no significance finding *anti*-particles. It only proves the variety of the quark composition of hadrons.

The energy and mass change also can be shown:

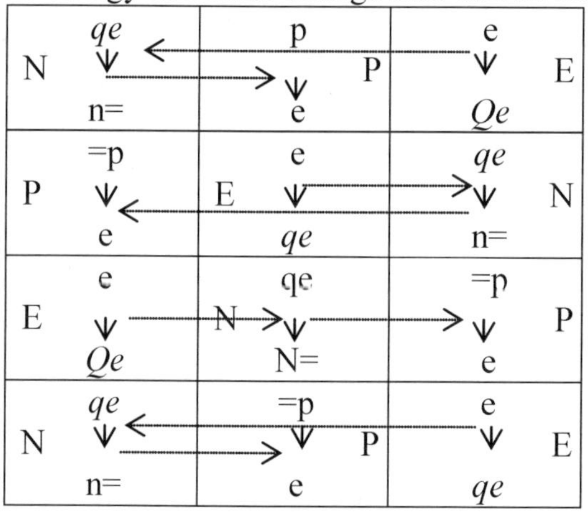

<table>
<tr><td>qe
N ∨ ←
n=</td><td>p
→ ∨ P
e</td><td>e
∨ E
Qe</td></tr>
<tr><td>=p
P ∨ ←
e</td><td>e
E ∨ →
qe</td><td>qe
∨ N
n=</td></tr>
<tr><td>e
E ∨ →
Qe</td><td>qe
N → ∨ →
N=</td><td>=p
∨ P
e</td></tr>
<tr><td>qe
N ∨ ←
n=</td><td>=p
→ ∨ P
e</td><td>e
∨ E
qe</td></tr>
</table>

The *blue shift* impact of the *electron process* drives the *neutron process* through *red shift*. The *neutron process*, the *collapse* of the *quantum entropy*, the re-transformation of energy into mass, *initiates* the *proton process*. The transformation of mass into energy is the *proton process* results in *electron* and the cycle starts again.

Fig. 25.4

Fig. 25.4

S. 25.3

25.3
Unbalanced *Strong* and *Weak Forces*
Nuclear chain reaction in nuclear reactors

The fission chain-reaction capability of ${}^{92}_{235}U$ isotopes, in certain and limited concentration within the natural ${}^{92}_{238}U$, is the result of its not balanced *Weak Interaction*, the not fully used *blue shift* impact of electrons by the neutron process. [The creation of *blue shift* surplus within an element in the *Nature* with more intensive neutron process ($Z > 1$) is a conflict between the *Weak* and the *Strong Interactions* of the element.] The standard – reference to the balanced ${}^{92}_{238}U$ – *blue shift* impact – is used not in full in ${}^{92}_{235}U$.

One part of the *Strong Interactions* of the ${}^{92}_{235}U$ element (the proton process) creates sufficient electron process and *Weak Force* – while the other part (the neutron process) does not use it in full.

In other words, ${}^{92}_{235}U$ generates more energy than its internal demand. The neutron process is less intensive than it should be. This is the disharmony which creates the conflict: The electron *blue shift* surplus of ${}^{92}_{235}U$ creates *blue shift* conflict within the ${}^{92}_{235}U$ isotope. The proof of this conflict is the heat generation. The 0.7% share of ${}^{92}_{235}U$ isotopes within the ${}^{92}_{238}U$ element does not lead to big problems. *Nature* tolerates this deficiency with limited heat generation.

The response of ${}^{92}_{235}U$ isotopes to a higher concentration of their share within ${}^{92}_{238}U$ or to additional *blue shift* impacts from external source (water, carbon or other moderators with *blue shift* surplus) is their increased conflict – *fission*: ${}^{92}_{235}U$ cannot keep its elementary structure. The increased *Weak Force* (*blue shift* impact) creates increased internal *blue shift* conflict within the element and destroys it.

The higher the impact of the *Weak Force* is, the stronger is the effect of the imbalance between the proton and the neutron processes (its *Strong Interactions*) within the ${}^{92}_{235}U$ isotope: The consequence is controlled or uncontrolled nuclear reaction, with birth of new elements.

These new elements will all be with less intensive neutron process, with less *Z*, event-concentration value ($1.3 < Z < 1.4$), than $^{92}_{235}U$. The Z of $^{92}_{235}U$ is still above 1.5. It cannot be differently, since these newly born elements will all be of less mass and less energy intensity than the one being destroyed.

$^{92}_{235}U$ is falling apart, because of the destroying *Weak Force,* the increased *blue shift* impact. The newly born *Cadmium*, *Indium Tin*, *Antimony*, *Tellurium*, *Iodine*, *Xenon*, *Caesium*, *Barium* will leave significant *Strong Interaction* out of balance – because of their lower *Z* value. *Weak Force* initiates the collapse, but there is no energy generation (proton process) for covering the needs of the freed neutron collapse.

The continuity of the proton-neutron process simply fails to happen. As a consequence, the neutron collapse will swallow all available energy around.

Neutron radiation in fact is not other than energy portions taken from other elementary structures capable of providing energy in order to balance the mass collapse of the *Strong Interaction*: mainly from elements and molecules either with less intensive neutron process – like compounds with water or carbon, which at the same time also intensify the *blue shift* conflict and heat generation - or heavy metals, less sensitive, more tolerant to the proton-neuron energy and mass balance. (The second case is equivalent to $^{92}_{235}U$ isotopes of less acting concentration.)

In the case of an uncontrolled destruction – this *blue shift* demand of neutrons is equivalent to *Black Hole* effect.

Water flow in controlled chain reaction has two functions: (1) *moderator* as providing additional electron *blue shift* surplus, and, creating with that, additional conflict, resulting in the speeding up of the destruction; and (2) cooling, taking away heat, part of the conflict – if the increase of the *blue shift* conflict is slow enough. Should the water flow be stopped, the destructive effect of the imbalance will prevail. The destruction does not just create new elements, but also – because of the *energy need* of the free neutrons around – causes damages in elementary structures of other elements, resulting in *gamma*, *alpha* and *beta* radiation.

There is no mass transformation without energy transformation. The balance of *Strong* and *Weak Interaction* is the basis of *the continuity of matter*.

S. 26

26

Acceleration effect of rotation

Ref Fig. 26.1

We take a disc on Fig. 26.1 with infinite number of elementary mass points. Each spot of the disc is with electron process *blue shift* conflict (light signal). The *blue shift* conflict is an external impact to the *Quantum Membrane*. We examine the effect of the rotation of the disc on this impact.

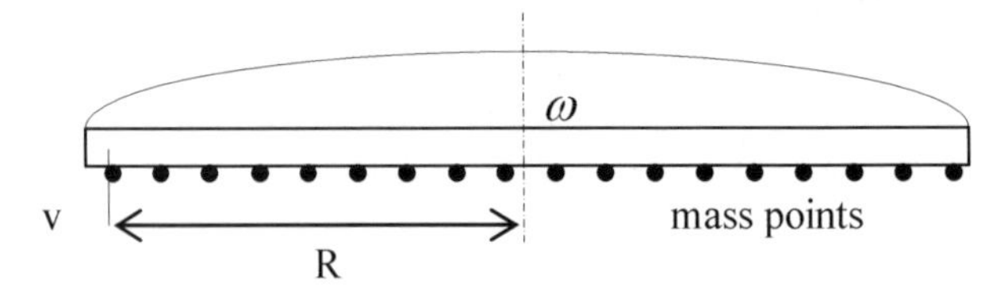

Fig 26.1

Fig.26.1

We take it that at the stationary status of the disc, the internal elementary impact to the *Quantum Membrane* within each spot of the disc is equal to:

26A1

$$w = \frac{dmc^2}{dt_o}\sqrt{1-\frac{i^2}{c^2}}\left(1-\sqrt{1-\frac{(c-i)^2}{c^2}}\right) = f_i \cdot q$$

The full description of the internal elementary impact at stationary status, with time flow and intensity values of $dt_i = 1$ and $\varepsilon_i = 1$ is:

26A2 26A3

$$\frac{dmc^2}{dt_i\varepsilon_i}\left(1-\sqrt{1-\frac{(c-i)^2}{c^2}}\right) = \frac{dn}{dt_i\varepsilon_i}q; \quad \text{or} \quad \dot{m}_i c^2\left(1-\sqrt{1-\frac{(c-i)^2}{c^2}}\right) = f_i \cdot q$$

With the rotation of the disc the impact of the *blue shift* conflict – equal in each spots at the stationary status of the disc – is changing. The impact depends on *R*, the *radius* of the spot from the centre of the disc and ω the angular speed of the rotation.

Taking the angular speed for constant value, the elementary mass spots will have peripheral speed:

$$v = \omega \cdot R \qquad \text{26A4}$$

The change of the peripheral speed by the radius of the disc is: $\frac{dv}{dR} = \omega$ 26A5

With reference to Section 4, rotation is a complex motion, with its peripheral and radial (*virtual* in the direction of the radius of the rotation) acceleration in effect.
The virtual speed of mass points of the rotating disc in the direction of the radius is:

$$\upsilon = \frac{dR_\upsilon}{dt_\upsilon} \quad \text{where } dt_\upsilon = \frac{dt_v}{\sqrt{1-(\upsilon^2/c^2)}}; \quad \text{and} \quad dR_\upsilon = \frac{dR_v}{\sqrt{1-(\upsilon^2/c^2)}}; \qquad \text{26B1, 26B2}$$

It is easy to accept that the *speed* of the motion (as the peripheral so is the virtual-radial) measured within the stationary systems of reference and within the systems of reference of the peripheral and the radial motion are equal:

$$\upsilon = \frac{dR_\upsilon}{dt_\upsilon} = \frac{dR_v}{dt_v} = \frac{dR_o}{dt_o} \quad \text{or} \quad \upsilon = R_\upsilon n_u = R_v n_v = R_o n_o \qquad \text{26B3}$$

For a single mass point the formula in 26B1 is the speed of the motion in the direction of the radius. For a rigid mass point – part of the rotating disc – this speed is *virtual*, but the centrifugal effect is real and acting. (The difference is in fact in electron *blue shift*: either conflicting or not.) The relation of the peripheral speed and the virtual speed of any of the infinite number of mass points is:

$$v = 2\Pi\frac{dR}{dt} \quad \text{and} \quad \upsilon = \frac{v}{2\Pi} \qquad \text{26B4}$$

The angular speed is: $\omega = \frac{2\Pi}{t} = 2\Pi n$; The peripheral speed is: $v = \frac{2\Pi R}{t} = 2\Pi Rn$ 26B5

The virtual speed of the motion in the direction of the radius is: $\upsilon = \frac{v}{2\Pi} = Rn$ 26B6

where the time of a single spin: $t = \frac{1}{n}; \left[\frac{1}{\sec}\right]$. Since: $t = \frac{2\Pi}{\omega} = \frac{1}{n}$ therefore $n = \frac{\omega}{2\Pi}$

There could be two kinds of acceleration identified in this motion:

1 – *the acceleration of the peripheral speed* $a_v = \frac{dv}{dt_v} = \frac{d\omega_v}{dt_v}R_v + \frac{dR_v}{dt_v}\omega_v$ 26C1

The definition of this acceleration also includes the acceleration in the direction of the radius. (The physical meaning of this acceleration is the effect of the permanent change in the direction of the motion.)

At a given and constant angular speed, it is: $\frac{d\omega_v}{dt_v} = 0$ and $a_v = \frac{dv}{dt_v} = \frac{dR_v}{dt_v}\omega_v$ 26C2

v_v ω_v t_v R_v n_v are measured within the system of reference of the peripheral motion.

From 25B1 and 25B5 follows that the value of this acceleration is:

26C3 $$a_v^v = \frac{dv}{dt_v} = \frac{dR}{dt_v}\omega_v = \upsilon \cdot \omega_v = \frac{v}{2\Pi}\omega_v = \frac{2\Pi R_v}{2\Pi} n_v 2\Pi n_v = 2\Pi R_v n_v \cdot n_v = v \cdot n_v$$

This acceleration means permanent change of the direction of the motion.
The acceleration measured within the stationary system of reference is:

26C4 $$a_o^v = 2\Pi R_o n_o \cdot n_o = v \cdot n_o \qquad \text{since} \quad dt_v = \frac{dt_o}{\sqrt{1-(v^2/c^2)}}; \quad \text{and} \quad n_v = n_o\sqrt{1-(v^2/c^2)}$$

The relation between the values of the acceleration, measured within the stationary and the peripheral systems of reference is:

26C5 $$a_v^v = v \cdot n_v = v \cdot n_o\sqrt{1-(v^2/c^2)} = a_o^v\sqrt{1-(v^2/c^2)}$$

26C6 2 – *the acceleration within the direction of the radius*: $$a_\upsilon^\upsilon = \frac{d\upsilon}{dt_\upsilon} = Rn \cdot n_\upsilon = \upsilon \cdot n_\upsilon$$

where $$dt_\upsilon = \frac{dt_v}{\sqrt{1-(\upsilon^2/c^2)}}; \quad \text{and} \quad n_\upsilon = n_v\sqrt{1-(\upsilon^2/c^2)}$$

The physical meaning of the radial acceleration is the effect of the virtual motion in the direction of the growing radius. (Reference to the system of reference of the peripheral motion.) The radial acceleration, measured from the stationary system of reference is:

26C7 $$a_o^\upsilon = \frac{d\upsilon}{dt_o} = \upsilon \cdot n_o$$

26C8 The relation is: $$a_\upsilon^\upsilon = \upsilon \cdot n_\upsilon = \upsilon \cdot n_o\sqrt{1-(v^2/c^2)}\sqrt{1-(\upsilon^2/c^2)} = a_o^\upsilon\sqrt{1-(v^2/c^2)}\sqrt{1-(\upsilon^2/c^2)}$$

The mass points of the (rigid) rotating disc are under permanent effect of acceleration in the direction of the radius. The accelerating effect depends on the speed of the rotation and the radius.

Ref S.20 The solid structure of the rotating disc (with $Z>1$) prevents mass particles from motion in the direction of the radius. It results, however, in change of intensity values and time systems of events within the elementary structure of the rotating disc.

If the source of the motion of the mass spots is the (natural event of the) transformation of mass into energy, there is no accelerating force and no centrifugal effect. [The sphere symmetrical expanding acceleration (proton process) and accelerating collapse (neutron process) are natural events, consequence of the internal mass energy balance of the matter. The electron process is event with constant speed and permanent acceleration.]

If the acceleration is generated by external energy, the source of the acceleration (result of rotation) is different than the natural transformation of mass into energy. The mass-energy balance of the spots will be disrupted. This conflict generates accelerating *centrifugal force* and impact. Should the mass points at the spot of the disc be free, they would be accelerating in the direction of the growing radius.

The status of all mass spots of the structure of the disc corresponds to status of rest. This de facto acts against the motion. The reason is the conflict of intensities and mass values of the fixed spots of the rigid rotating disc. Without external work the rotation stops.

Permanent external work results in acceleration without change of the mass but in permanent change of the intensity of the system of reference of the acceleration. The accelerating impact depends on the value of the radius and the angular speed (or speed) of the rotation.

The rotation is similar to the electron process: the speed is constant and the acceleration is permanent – resulting in work and resolving the conflict of the "constant" mass of the particles and the intensity of the event.

We can identify two kinds of speed and acceleration, as above presented. In the case of peripheral speed, the acceleration means permanent change in the direction. In the case of radial speed, the acceleration and the centrifugal force are real, the motion is virtual and the speed is constant. The acceleration happens within the system of reference with constant speed, perpendicular to the direction of the motion.

26.1 Intensity of events - acceleration

S 26.1

We take mass m and characterise its acceleration from the point of view of two systems of reference. The classical description of the force formula, the effect of the rotation, is:

$$F = m \cdot d\left(\frac{ds}{dt}\right)\frac{1}{dt} \qquad \text{26D1}$$

The full and comprehensive description of 26D1 is: $F = \frac{dm}{dt\varepsilon} d\left(\frac{dv}{dt}\right)\frac{1}{dt\varepsilon}$ 26D2

For the explanation of 26D2 we have to note that mass is only measurable through its effect. In conventional meaning:

$\frac{dm}{dt} = \dot{m}$ and intensities are not part of the description.

The same mass of the same acceleration has the same equal absolute force effect, even if observed from two different systems of reference:

$$\frac{dm}{dt_o\varepsilon_o} d\left(\frac{ds_o}{dt_o}\right)\frac{1}{dt_o\varepsilon_o} = \frac{dm}{dt_v\varepsilon_v} d\left(\frac{ds_v}{dt_v}\right)\frac{1}{dt_v\varepsilon_v} \qquad \text{26D3}$$

Since $dt = \frac{1}{\varepsilon}$ and the speed value $v = \frac{ds_o}{dt_o} = \frac{ds_v}{dt_v} \ldots = \frac{ds_n}{dt_n}$, 26D3 gives perfect equality of the effect or impact in absolute terms in any system of reference of measurement!

(The equality here is all right, since time and intensity values relate to the same rotation.)
The system of reference of the observation on the left hand side of 26D3 is the stationary one, on the right hand side is the one in motion

The mass and the speed keep changing simultaneously and the equality in 26D3 is no more valid with the intensities taken out from the equation:

26D4 1: $$\frac{dm}{dt_o} d\left(\frac{ds_o}{dt_o}\right)\frac{1}{dt_o\varepsilon_o} \neq \frac{dm}{dt_v} d\left(\frac{ds_v}{dt_v}\right)\frac{1}{dt_v\varepsilon_v}$$ - the mass effect is without intensity note

- the impacts of the same and perfectly equal masses on the two sides of the 26D4 are *different*: the measured mass values within the two different systems of reference are different: $\frac{dm}{dt_o} = \dot{m}_o \neq \dot{m}_v = \frac{dm}{dt_v}$

26D5 2: $$\frac{dm}{dt_o} d\left(\frac{ds_o}{dt_o}\right)\frac{1}{dt_o} \neq \frac{dm}{dt_v} d\left(\frac{ds_v}{dt_v}\right)\frac{1}{dt_v}$$ - also the acceleration is without intensity note

- the change of the equal speed on both sides of 26D5 may have different intensities: the acceleration of the masses is different.

Within a stationary system of reference, where relative conditions of $dt_o = 1$ and $\varepsilon_o = 1$ are taken, the absolute value of the impact (force) is equal to its relativistic value. This condition, however, is not valid any more in other systems of reference, different in motion: $dt_o \neq dt_v$; $\varepsilon_o \neq \varepsilon_v$.
The intensity of the mass change has its certain value in systems of reference. It might be noted, might be not. Without comparison of events happening within systems of reference different in motion – the missing intensity within the formulas has no consequence.

26D6 $$\frac{dmc^2}{dt \cdot \varepsilon} = \frac{dn}{dt \cdot \varepsilon} q; \quad \text{or} \quad \dot{m}c^2 = f \cdot q$$

26D6 means a certain mass change corresponds to a certain quantum impact, frequency. This formula relates and identifies the same system of reference on both sides. In a different system of reference the effect in 26D6 would be different.

The description of the work, the acceleration of the mass in 26E1 below identifies the system of reference through its time and intensity parameters:

26E1 $$\frac{dmc^2}{dt \cdot \varepsilon}\left(1 - \sqrt{1 - \frac{v^2}{c^2}}\right) = \frac{dW}{dt \cdot \varepsilon}; \quad \text{or} \quad \dot{m}c^2\left(1 - \sqrt{1 - \frac{v^2}{c^2}}\right) = \dot{w} = \Delta f \cdot q$$ as in 26D6,

– but it does not characterise the intensity of the process itself!

With reference to Section 20 and 24, elements differ in the *relation* of their proton and neutron process intensities. The distinguishing key is the electron process, the acceleration (intensity of the mass change) for the same, quasi constant speed difference of $\Delta v = (c - i)$:

$$\varepsilon = \frac{(c-i)}{\Delta t} = a$$

The system of reference of the electron process for all elements is the same: event at speed $i = \lim a\Delta t = c$. The difference can only be described through the intensity of the process, as indicator of the event:

$$d\frac{dmc^2}{dt_i\varepsilon_i}\left(1-\sqrt{1-\frac{(c-i)^2}{c^2}}\right)\frac{1}{dt_i\varepsilon_e} = W \qquad \text{26E2}$$

In other words, the electron-process-event of all elements happens in the same time system, but with different intensities.

26E2 in simplified description is:

$$\frac{dmc^2}{dt_i\varepsilon_x}\left(1-\sqrt{1-\frac{(c-i)^2}{c^2}}\right) = W \qquad \text{26E3}$$

where the time and the intensity with index i (dt_i and ε_i) denote the system of reference of the electron or the electron process, equal for all elements. Intensity values with index e and x (ε_e and ε_x) characterise the electron process of a certain element and ε_x is the integrated intensity of the element, value of: $\varepsilon_x = \varepsilon_i \cdot \varepsilon_e$ 26E4

The description of the electron process of the elementary structure of the disc at stationary status is different than in rotation.

- The full description at the stationary status is:

$$d\left[d\left(\frac{dmc^2}{dt_i\varepsilon_i}\left(1-\sqrt{1-\frac{(c-i)^2}{c^2}}\right)\frac{1}{dt_i\varepsilon_e}\right)\frac{1}{dt_o\varepsilon_o}\right] = dd\frac{dn}{dt_i\varepsilon_i}q\frac{1}{dt_i\varepsilon_e}\frac{1}{dt_o\varepsilon_o} \qquad \text{26E5}$$

26E5 can be written in a more simplified form:

$$d\frac{dmc^2}{dt_i\varepsilon_x}\left(1-\sqrt{1-\frac{(c-i)^2}{c^2}}\right)\frac{1}{dt_o\varepsilon_o} = d\frac{dn}{dt_i\varepsilon_x}q\frac{1}{dt_o\varepsilon_o};$$

26E6

or

$$\frac{d[Event]}{dt_o\varepsilon_o} = d\frac{dn}{dt_i\varepsilon_x}q\frac{1}{dt_o\varepsilon_o};$$

- The electron process in 26E5 and 26E6 can also be written as process impacted by the system of reference of the rotation. Since the rotation is the result of external energy, different than the electron process itself, ε_R in the formula will characterise this difference:

26E7 $$\frac{d[Event]}{dt_o\varepsilon_R} = d\frac{dn}{dt_i\varepsilon_x} q\frac{1}{dt_o\varepsilon_R}$$

26E8 26E6 and 26E7 are obviously different: $$\frac{d[Event]}{dt_o\varepsilon_o} \neq \frac{d[Event]}{dt_o\varepsilon_R}$$

For making 26E8 equal, it shall be adjusted.
The way of this adjustment is important for the assessment of the impact. There are two options to do it:

26F1 $$\frac{d[Event]}{dt_o\varepsilon_R} \nearrow = \Phi\frac{d[Event]}{dt_o\varepsilon_o}$$ - means more of the same event – (but the same intensity)

26F2 $$\frac{d[Event]}{dt_o\varepsilon_R} \searrow = \frac{d[\Phi \cdot Event]}{dt_o\varepsilon_o}$$ - means different event – (different intensity)

In the case of the *electron process* in acceleration a "different event" cannot be an option. Different event would mean different intensity, different electron process (and different element). The only option is the *blue shift* of the same intensity,

26F3 26F1 gives: $\varepsilon_o = \Phi_e \cdot \varepsilon_R$ and $\Phi_e = \varepsilon_o / \varepsilon_R$

26F4 $\varepsilon_R > \varepsilon_o$ means: there is an electron process deficit indeed, since: $\Phi_e < 1$

At stationary status of the disc $\varepsilon_o = 1$ relates to dt_o, the system of reference of the measurement.

The intensity of the electron process is constant and cannot follow the dictated by the radial acceleration intensity grow of rotation!

In the case of *proton and neutron processes*, where the intensity of both can change simultaneously, the description would be different, the one in 26F2.

$\varepsilon_R > \varepsilon_o$ means: the proton and neutron processes are different – increased intensity of mass change:

26F5 $$\frac{d[\Phi Event]}{dt_o} > \frac{d[Event]}{dt_o} \text{ intensity increase } \rightarrow \frac{d(\Phi m)}{dt_o} > \frac{dm}{dt_o}$$

The assessment has also to consider the impact of the peripheral speed.

>> Peripheral impact results in two different statuses of the acceleration of the *electron process*:

$$\frac{d[Event]}{dt_o\varepsilon_{Rv}} - \frac{d[Event]}{dt_o\varepsilon_R} = \Phi_e \frac{d[Event]}{dt_o\varepsilon_o}\left(\frac{1}{\sqrt{1-(v^2/c^{2)}}} - 1\right); \quad \text{and } \Phi_e < 1 \qquad \text{26F6}$$

$$\frac{d[Event]}{dt_o\varepsilon_{Rv}} = \Phi_e \frac{d[Event]}{dt_o\varepsilon_o\sqrt{1-(v^2/c^2)}}; \quad \varepsilon_{Rv} = \varepsilon_R\sqrt{1-(v^2/c^2)}$$

- (1) *electron process deficit* as a result of the increase of the peripheral speed further increases:

$$\frac{\varepsilon_o}{\varepsilon_R\sqrt{1-(v^2/c^2)}} = \frac{\varepsilon_o}{\varepsilon_{Rv}} = \frac{\Phi_e}{\sqrt{1-(v^2/c^2)}} = \Phi_{ev} \quad \text{until} \quad \frac{1}{\varepsilon_R\sqrt{1-(v^2/c^2)}} = 1; \qquad \text{26F7}$$

$\varepsilon_o = 1$ means, the intensity of the electron (and other) processes corresponds to its standard value – without rotating impact. With reference to 26B4, 26C7 and 26B6, the intensity of the radial acceleration is: $\varepsilon_R = a_o^\upsilon = \upsilon \cdot n_o = \frac{v}{2\Pi} n_o = R \cdot n_o^2$

- (2) but since the speed increase of the motion, above this peripheral speed limit slows down the time flow, *the intensity deficit disappears and creates surplus.*

$$v \geq c\sqrt{1 - \frac{1}{\varepsilon_R^2}} \qquad \text{26F8}$$

The higher is $\varepsilon_R = R \cdot n_o^2$ the intensity, the higher is this speed limit.

The closer is the mass spot on the disc to the centre of the rotation, the higher is the *blue shift* surplus of the electron process!

>> The peripheral speed for the proton and neutron processes will further increase the intensity:

$$\frac{d[\Phi Event]}{dt_o\varepsilon_{Rv}} - \frac{d[\Phi Event]}{dt_o\varepsilon_R} = \frac{1}{dt_o\varepsilon_R}\frac{d[\Phi Event]}{\sqrt{1-(v^2/c^2)}} - \frac{d[\Phi Event]}{dt_o\varepsilon_R} \qquad \text{26F9}$$

26F9 demonstrates why rotation needs work and how the centrifugal force is acting.

The difference in the behaviours of the proton and neutron versus electron processes is that the electron process is of constant intensity.

26.1.1. The intensity values of the change   S. 26.1.1

We measure *work-intensities and energy-intensities* in systems of reference rather than absolute values.

With reference to 26C1 – 26C8 the intensity values of the rotation are the following:

> *Intensity* of the permanent change of the direction of the peripheral motion:

26G1
$$\varepsilon_o^v = \frac{dv}{dt_o} = a_o^v = \frac{a_v}{\sqrt{1-(v^2/c^2)}} = \frac{\varepsilon_v}{\sqrt{1-(v^2/c^2)}}; \quad \text{and} \quad \varepsilon_v = \frac{dv}{dt_v} = a_v$$

> *Intensity* of the change of the virtual radial motion:

26G2
$$\varepsilon_o^\upsilon = \frac{d\upsilon}{dt_o} = a_o^\upsilon = \frac{a_\upsilon}{\sqrt{1-(v^2/c^2)}\sqrt{1-(\upsilon^2/c^2)}} = \frac{\varepsilon_\upsilon}{\sqrt{1-(v^2/c^2)}\sqrt{1-(\upsilon^2/c^2)}}$$

26G3
$$\text{and} \quad \varepsilon_v^\upsilon = \frac{d\upsilon}{dt_v} = a_v^\upsilon = \frac{a_\upsilon}{\sqrt{1-(\upsilon^2/c^2)}} = \frac{\varepsilon_\upsilon}{\sqrt{1-(\upsilon^2/c^2)}} \quad \text{and} \quad \varepsilon_\upsilon = \frac{d\upsilon}{dt_\upsilon} = a_\upsilon$$

S 26.2

26.2
Time shift as result of the rotation

The rotating disc generates time shift. The intensity change of the proton and neutron processes results in the change of the duration of the processes:

26H1
$$\frac{dmc^2}{\left(\frac{dt_p}{x}\right)x\varepsilon_p}\left(1-\sqrt{1-\frac{i^2}{c^2}}\right) = \frac{dmc^2}{\left(\frac{dt_n}{x}\right)x\varepsilon_n}\sqrt{1-\frac{(c-i)^2}{c^2}}\left(\sqrt{1-\frac{i^2}{c^2}}-1\right)$$

x in 26H1 denotes the increase of the intensity, result of the acceleration, and $x > 1$

Since the intensity of the electron processes cannot be changed, this results in shift between the duration of the proton-neutron versus electron processes.

Acceleration as external impact influences the elementary process.
For explaining this, we assess the motion at constant speed and acceleration

<u>*Motion with constant speed v*</u>:
We take a proton process of an element in linear motion with constant speed (no acceleration) in absolute terms and intensities and assess the impact of the motion:

26I1
26I2
$$\frac{dmc^2}{dt_p\varepsilon_p}\left(1-\sqrt{1-\frac{i^2}{c^2}}\right); \quad \text{and} \quad \dot{m}_p c^2\left(1-\sqrt{1-\frac{i^2}{c^2}}\right) = \frac{dn}{dt_p}q = f_p \cdot q$$

26I3
Blue shift during the motion with v: $$\frac{dmc^2}{dt_p\varepsilon_p}\left(1-\sqrt{1-\frac{i^2}{c^2}}\right)\left(2-\sqrt{1-\frac{v^2}{c^2}}\right);$$

"p" in the index means: process

Red shift during the motion with *v*: $$\frac{dmc^2}{dt_p\varepsilon_p}\left(1-\sqrt{1-\frac{i^2}{c^2}}\right)\sqrt{1-\frac{v^2}{c^2}} \qquad 26I4$$

The summarised effect of the *blue* and *red shift* of the motion with *v* is:

$$2\frac{dmc^2}{dt_p\varepsilon_p}\left(1-\sqrt{1-\frac{i^2}{c^2}}\right)\frac{1}{2}=\frac{dmc^2}{dt_p\varepsilon_p}\left(1-\sqrt{1-\frac{i^2}{c^2}}\right) \qquad 26I5$$

In the case of multiply (acting in parallel) *blue* and *red shift* impacts:

$$X\frac{dmc^2}{dt_p\varepsilon_p}\left(1-\sqrt{1-\frac{i^2}{c^2}}\right)\frac{1}{X}=\frac{dmc^2}{dt_p\varepsilon_p}\left(1-\sqrt{1-\frac{i^2}{c^2}}\right) \qquad 26I6$$

26I6 in intensity terms is $$\dot{m}_pc^2\left(1-\sqrt{1-\frac{i^2}{c^2}}\right)=\frac{dn}{dt_p}q=f_p\cdot q \qquad 26I7$$

and with reference to 26I2 $$f_p=f_p \qquad 26I8$$

Acceleration with a – the consequence is different:
The *blue* and the *red shifts* during acceleration with *a*, resulting intensity ε_a.

Blue shift, result of acceleration: $$\frac{d}{dt_a\varepsilon_a}\frac{dmc^2}{dt_p\varepsilon_p}\left(1-\sqrt{1-\frac{i^2}{c^2}}\right)\left(2-\sqrt{1-\frac{(a\Delta t)^2}{c^2}}\right);\text{ and} \qquad 26J1$$

Red shift, result of acceleration: $$\frac{d}{dt_a\varepsilon_a}\frac{dmc^2}{dt_p\varepsilon_p}\left(1-\sqrt{1-\frac{i^2}{c^2}}\right)\sqrt{1-\frac{(a\Delta t)^2}{c^2}} \qquad 26J2$$

The summarised effect is:

$$2\frac{d}{dt_a\varepsilon_a}\frac{dmc^2}{dt_p\varepsilon_p}\left(1-\sqrt{1-\frac{i^2}{c^2}}\right)\frac{1}{2}=\frac{d}{dt_a\varepsilon_a}\frac{dmc^2}{dt_p\varepsilon_p}\left(1-\sqrt{1-\frac{i^2}{c^2}}\right) \qquad 26J3$$

Or in the case of multiply parallel *blue-red shifts*:

$$X\frac{d}{dt_a\varepsilon_a}\frac{dmc^2}{dt_p\varepsilon_p}\left(1-\sqrt{1-\frac{i^2}{c^2}}\right)\frac{1}{X}=\frac{d}{dt_a\varepsilon_a}\frac{dmc^2}{dt_p\varepsilon_p}\left(1-\sqrt{1-\frac{i^2}{c^2}}\right) \qquad 26J4$$

The result is: $$\frac{dmc^2}{\frac{dt_p}{x}x\cdot\varepsilon_p}\left(1-\sqrt{1-\frac{i^2}{c^2}}\right) \qquad 26J5$$ or: $$\dot{m}_{pa}c^2\left(1-\sqrt{1-\frac{i^2}{c^2}}\right)=\frac{dn}{\frac{dt_p}{x}}q=f_{pa}\cdot q \qquad 26J6$$

and $$\dot{m}_{pa}c^2\left(1-\sqrt{1-\frac{i^2}{c^2}}\right)>>\dot{m}_pc^2\left(1-\sqrt{1-\frac{i^2}{c^2}}\right) \qquad f_{pa}>>f_p \qquad 26J7$$

External acceleration impacts the internal sphere symmetrical acceleration.
The external *blue* and *red shifts* impact the internal elementary proton and neutron processes. The internal process works against the external impact.

In line with the continuity law of matter, the proton and neutron processes are equally impacted by the intensity growth:

26K1 $\dfrac{\dfrac{dmc^2}{dt}}{\dfrac{\varepsilon \cdot x}{x}}[Event]$ and $\dot{m}_{event}c^2[Event]$ grow with the increase of the speed of the rotation

26K2 $\varepsilon_e = \dfrac{\varepsilon_{pa}}{\varepsilon_{na}}\sqrt{1-\dfrac{(c-i)^2}{c^2}}$ the main characteristic of the element is constant.

Since the electron process must keep its time and intensity values, the time difference results in *time shift* between the proton-neutron and the electron processes: The *blue shift* impact of the electron process works in time shift. With the increase of the acceleration ($a = 2\Pi Rn^2$ increase of speed or radius or both) the *time shift* grows.

The time count of the proton and the neutron processes will be modified by the peripheral speed of the rotation.

In the example of the neutron, the modification will be:

26K3 $$\frac{d[Event]}{dt_{nR}\varepsilon_{nR}} = \frac{d[Event]}{dt_{no}\varepsilon_{no}} = \frac{d[Event]}{\dfrac{dt_{no}}{a}a \cdot \varepsilon_{no}} = \frac{d[Event]}{dt_{no}\varepsilon_{no}} \quad \Rightarrow \quad \varepsilon_{nR} = a \cdot \varepsilon_{no}$$

26K4 $$\frac{d[Event]}{dt_{nR}\varepsilon_{nRv}} - \frac{d[Event]}{dt_{nR}\varepsilon_{nR}} = \frac{d[Event]}{dt_{nR}\varepsilon_{nR}}\left(\frac{1}{\sqrt{1-(v^2/c^2)}}-1\right); \quad \varepsilon_{nRv} = \varepsilon_{nR}\sqrt{1-(v^2/c^2)}$$

26K5 and $$\frac{d[Event]}{dt_{nR}\varepsilon_{nR}} = \frac{d[Event]}{dt_{nR}\varepsilon_{nRv}}\sqrt{1-\frac{v^2}{c^2}} = \frac{d[Event]}{dt_{nRv}\varepsilon_{nRv}} = \frac{d[Event]}{dt_{no}\varepsilon_{no}}$$

26K6 $dt_{nR} = \dfrac{dt_o}{a}$; and $dt_{nRv} = \dfrac{dt_o}{a\sqrt{1-(v^2/c^2)}}$

Since the *proton* and *neutron* processes are of similar nature, the time systems of the neutron and the proton processes are modified, increased by the peripheral speed.

The increased proton and neutron process intensities need more *blue shift*, more electrons. At the same time, the *time shift* of the electron process results in longer electron *blue shift* effect.

In the case of constant speed of the rotation, the gradient of the change of the intensity of the processes, as consequence of the radius is:

$$\frac{d\dot{m}_{pa}}{dR} = \frac{d\dot{m}_{na}}{dR} = f(a) \quad \frac{d\dot{m}_e}{dR} = 0 \tag{26L1}$$

Therefore:

electron process with time shift

$$\dot{m}_{pa}c^2\left(1-\sqrt{1-\frac{i^2}{c^2}}\right) >> \dot{m}_e c^2\sqrt{1-\frac{i^2}{c^2}}\left(1-\sqrt{1-\frac{(c-i)^2}{c^2}}\right) << \dot{m}_{na}c^2\sqrt{1-\frac{(c-i)^2}{c^2}}\left(\sqrt{1-\frac{i^2}{c^2}}-1\right) \tag{26L2}$$

intensified proton process

intensified neutron process

S. 27

27

Rotating Disc Quantum Device

Ref S.25

With reference to Section 26, rotation creates *blue shift* deficit or *blue shift* demand and electron process *time shift*.

The electron process of the elementary structure of the disc at stationary status, with reference to 25E6, can be written as

Ref 26E6

$$d\frac{dmc^2}{dt_i\varepsilon_x}\left(1-\sqrt{1-\frac{(c-i)^2}{c^2}}\right)\frac{1}{dt_o\varepsilon_o}=d\frac{dn}{dt_i\varepsilon_x}q\frac{1}{dt_o\varepsilon_o};$$

or

$$\frac{d[Event]}{dt_o\varepsilon_o}=\frac{\dot{e}_o}{\varepsilon_o}=d\frac{dn}{dt_i\varepsilon_x}q\frac{1}{dt_o\varepsilon_o};$$

where ε_x – characterises the intensity of the electron process of the elementary structure of the disc; dt_i – relates to the electron process, dt_o and ε_o – denote the stationary status.

The electron process within the rotating disc, at a certain radius, with reference to 26E7, is:

Ref 26E7

$$\frac{d[Event]}{dt_o\varepsilon_{Rv}}=\frac{\dot{e}_o}{\varepsilon_{Rv}}=d\frac{dn}{dt_i\varepsilon_x}q\frac{1}{dt_o\varepsilon_{Rv}}$$

Ref 26F3 26F6

The formulas in 26E6 and 26E7 represent the same electron processes, but because of the difference in the circumstances, they are of different absolute value – result of the rotation of the disc (driven by external energy): The intensities are different. Their relation, with reference to 26F3 and 26F6, is:

27A1

$$\frac{\varepsilon_{Rv}}{\varepsilon_o}=\frac{a_R}{a_o}\frac{1}{\Phi_e\sqrt{1-(v^2/c^2)}}$$

ε_{Rv} – denotes the intensity at the spot at radius R with angular speed of v

27A1 shows that with the increase of the radius and the speed of the rotation, a_R the radial acceleration and v the angular speed is permanently growing:

$a_R = R_o n_o^2$ and $v = 2\Pi R_o n_o$

R_o and n_o both measured within the stationary system of reference.

The neutron and proton processes are also impacted by the acceleration of the rotation, but whatever is the change of the intensity of the proton and neutron processes, the continuity law of matter (appearing as *Strong Force* in protons and neutrons) keeps the elementary structure unchanged.

In line with the *general rule of the continuity of the matter* – the relation of the proton and neutron processes is not changing:

$$d\frac{dmc^2}{dt_p \varepsilon_p}\left(1-\sqrt{1-\frac{i^2}{c^2}}\right)\frac{1}{dt_o \varepsilon_R} = d\frac{dmc^2}{dt_n \varepsilon_n}\sqrt{1-\frac{(c-i)^2}{c^2}}\left(\sqrt{1-\frac{i^2}{c^2}}-1\right)\frac{1}{dt_o \varepsilon_R}$$

Ref 19D2 27A2

27A2 can be written as

$$\frac{\dot{m}_p c^2}{\varepsilon_{pR}}\left(1-\sqrt{1-\frac{i^2}{c^2}}\right) = \frac{\dot{m}_n c^2}{\varepsilon_{nR}}\sqrt{1-\frac{(c-i)^2}{c^2}}\left(\sqrt{1-\frac{i^2}{c^2}}-1\right);$$

Ref 19E2

The proton and neutron processes are also intensified relative to the "stationary status", but their relation is without change and with reference to Sections 18 and 19, the intensity of the electron process shall remain unchanged, since the element remains unchanged:

$$\varepsilon_{eR} = \frac{\varepsilon_{pR}}{\varepsilon_{nR}}\sqrt{1-\frac{(c-i)^2}{c^2}} = \varepsilon_e \qquad 27A3$$

As consequence of the rotation, elementary processes go with *blue shift demand or deficit* and electron process *time shift*.

The increase in intensities of the neutron and proton processes, consequence of the rotation, generated by external work – in conventional terms means *centrifugal force*. The neutron process in solid objects by nature is more intensive than the proton process. (For the majority of the elements $Z>1$.) While there is no change in the quality of the elementary structure, as consequence of the rotation, this intensity increase explains the increase of the effect of the mass change (weigh). The relation of the proton and neutron processes remains unchanged, but the neutron process distinguishes *itself* by the *effect* of the more intensive mass change (resulting in centrifugal force).

!!!

27.1 Rotation *without* external electron flow

S. 27.1

As result of the accelerating effect of the rotation, the relation of the intensities of the proton and neutron processes is without change:

27B1 The relation of mass gradient values $\dot{m}_p$ and $\dot{m}_n$ are: $\frac{\dot{m}_p}{\dot{m}_n} = \sqrt{1 - \frac{(c-i)^2}{c^2}}$

The demand in electron process effect however is increased:

27B2

$$\Phi \cdot e_e = \Phi \cdot \dot{m}c^2 \sqrt{1 - \frac{i^2}{c^2}} \left(1 - \sqrt{1 - \frac{(c-i)^2}{c^2}} \right)$$

Ref 26C3 26C4

The gradient of the neutron and the proton mass change by the radius of the rotation is:

27B3

$$\frac{d\dot{m}_n}{dR} = \frac{d\varepsilon_n}{dR} = \frac{da}{dR} = \frac{2\Pi n^2 dR}{dR} = 2\Pi n^2 = \frac{d\dot{m}_p}{dR} = \frac{d\varepsilon_p}{dR}$$

The relation of intensities means:
the duration of the event is reciprocal to their intensities:

27B4

$$\varepsilon_e = \frac{\varepsilon_p}{\varepsilon_n} \sqrt{1 - \frac{(c-i)^2}{c^2}} \quad \text{and} \quad dt_e \sqrt{1 - \frac{(c-i)^2}{c^2}} = \frac{dt_n}{dt_p}$$

More intensive proton than neutron process means: proton process happens for shorter time period and the *blue shift* of electrons available can be used simultaneously for providing *red shift* to other elements.
(These are the chemically active elements.)

In the case of neutron intensive elements, the balance is also granted, but the neutron process can use "external" electron *blue shift* from others. The duration of the neutron process is shorter in time than the proton process and shorter than the electron provision of the elementary structure.

The time system of the neutron and the proton processes are different.

If we want to illustrate it with the same event

Duration

27C1 ———————— P $dt_p \cdot \varepsilon_p = 1$ $dt_p < dt_n$ and

———————————— N $dt_n \cdot \varepsilon_n = 1$ $\varepsilon_p > \varepsilon_n$

27C2 ———————————— P $dt_p > dt_n$ and

———————— N $\varepsilon_p < \varepsilon_n$

The electron process is the one, which connects the processes of the two hadrons: electron *generation* relates to the intensity of the proton process; the utilisation of the *blue shift* (neutron process) starts from *quantum entropy* stage.

Having the intensity of the elementary process increased in x times of the normal – result of acceleration – we have to note that in the case of

- proton intensive elements: more *blue shift* is available than in absolute terms used to be;
- neutron intensive elements: more *blue shift* is needed than in absolute terms are available.

While in absolute terms the equality is granted:

$$\frac{dmc^2}{dt_p\varepsilon_p}\left(1-\frac{v^2}{c^2}\right)=\frac{dmc^2}{dt_n\varepsilon_n}\sqrt{1-\frac{(c-i)^2}{c^2}}\left(1-\frac{v^2}{c^2}\right) \qquad \text{27C3}$$

Intensity values of the neutron and proton processes are different:

$$\frac{dmc^2}{dt_p}\left(1-\frac{v^2}{c^2}\right)\neq\frac{dmc^2}{dt_n}\sqrt{1-\frac{(c-i)^2}{c^2}}\left(1-\frac{v^2}{c^2}\right) \qquad \text{27C4}$$

or establishing conditional equality:

$$Z\frac{dmc^2}{dt_p}\left(1-\frac{v^2}{c^2}\right)=\frac{dmc^2}{dt_n}\sqrt{1-\frac{(c-i)^2}{c^2}}\left(1-\frac{v^2}{c^2}\right); \qquad Z=\frac{\varepsilon_n}{\varepsilon_p}=\frac{N}{P} \qquad \begin{matrix}\text{27C5}\\ \text{27C6}\end{matrix}$$

For standard elementary processes, the electron process is always in balance with the intensity relation of the proton-neutron process. This is the characteristic of the element. This is the reason why *Hydrogen* is *Hydrogen* and *Oxygen* is *Oxygen* etc. In the particular case of these two elements, the proton dominance is the reason why these elements are so active – they have electron process dominance and active electron process, especially *Hydrogen*.

In the case of external acceleration, the duration of the proton and the neutron processes is shortening. It leads to more intensive electron generation, meaning: the number of electron processes running parallel will be increased in line with the acceleration effect. The duration of the electron process however, reaching status of *quantum entropy* remains the same!

And the neutron process can start only after the *quantum entropy* stage – with the increased by the acceleration intensity.

The meaning of non equality conditions in 27C5 is without change

$$Z\cdot x\cdot\frac{dmc^2}{dt_p}\left(1-\frac{v^2}{c^2}\right)=x\cdot\frac{dmc^2}{dt_n}\sqrt{1-\frac{(c-i)^2}{c^2}}\left(1-\frac{v^2}{c^2}\right) \qquad \text{27D1}$$

27D2 which is in fact: $$Z\frac{\frac{dmc^2}{dt_p}}{x}\left(1-\frac{v^2}{c^2}\right)=\frac{\frac{dmc^2}{dt_n}}{x}\sqrt{1-\frac{(c-i)^2}{c^2}}\left(1-\frac{v^2}{c^2}\right)$$

and the electron process will be increased in numbers:

27D3 $$x\cdot e=x\cdot\frac{dmc^2}{dt_i}\left(1-\sqrt{1-\frac{(c-i)^2}{c^2}}\right)$$

there will be x times more electron processes, but the process time $dt_i = const$ remains the same.

>> In the case of *proton intensive elements*, where $Z<1$ – proton process provides more intensive electron generation than it is necessary, since

27D4 $$\frac{\frac{dmc^2}{dt_p}}{x}\left(1-\frac{v^2}{c^2}\right)>\underline{Z\frac{\frac{dmc^2}{dt_p}}{x}\left(1-\frac{v^2}{c^2}\right)}$$

intensity need of the acceleration

>> In the case of *neutron intensive elements*, where $Z>1$ – proton process provides less intensive electron generation than it is necessary, since

27D5 $$\frac{\frac{dmc^2}{dt_p}}{x}\left(1-\frac{v^2}{c^2}\right)<\underline{Z\frac{\frac{dmc^2}{dt_p}}{x}\left(1-\frac{v^2}{c^2}\right)}$$

intensity need of the acceleration

(Z is introduced in the formulas to visualise non-balance)

In absolute terms the mass balance is in order. In intensity terms, the imbalance has its significant consequences.

The other aspect of the effect of the acceleration is the *time shift* of the electron process. Electron generation is regulated by the proton process, electron *blue shift* is contributing to the neutron collapse, but the elementary process is controlled by *Strong Interaction* between hadrons. Meaning: whatever is the benefit or the loss in the intensity of the electron process in acceleration, the mass-energy transformation is the one (the proton and the neutron process) which determines the event:

At proton process dominance – the case is simple: additional electron *blue shift*, with all benefits to other processes.

At neutron process dominance – the electron process deficit is combined with electron process *time shift*: Neutron collapse goes with electron *blue shift* intensity deficit, but the constant duration of the electron process provides electron *blue shift* for long. This causes *time shift*. With the growth of the intensity this time shift also grows.

In a case of a solid rotating disc, the missing *blue shift* intensity on one side is compensated by electron process *time shift* on the other.

The generated electrons,

results of the proton process (in harmony with the neutron process) with standard dt_i electron process time, with the increased by the acceleration neutron process intensity and with decreased (dt_n / x) process time,

will be acting longer than the intensified neutron process lasts itself.

This provides additional *blue shift* reserve to later stages of the elementary process. The *time shift* with the acceleration grows.

With the rotation, the duration of the neutron process, as result of the intensity grow, will be less by:

$$\Delta t_a = \Delta t_{no} - \frac{\Delta t_{no}}{x} = \frac{\Delta t_{no}(x-1)}{x}; \; x = \frac{\varepsilon_{na}}{\varepsilon_{no}}; \; \Delta t_{na} = \frac{\Delta t_{no}}{x} \text{ and } dt_a = \frac{x-1}{x} dt_{no} \qquad 27E1$$

The gradient of the intensity change, reference to 27B2 is:

$$\frac{d\varepsilon_n}{dR} = x = 2\Pi n^2 \qquad 27E2$$

The shorter neutron process provides time reserve for the electron process: the electron process – with its standard, unchanged duration – will last longer.

If we conditionally divide the rotating disc into *n,* infinite number of elementary parts in the direction of the radius, spots, closer to the centre will always have time reserve to support the *red shift* demand of the elementary process at spots farer from the centre by the time reserve (surplus) of their electron *blue shift* duration.

The time shift of the electron *blue shift* covers part of the necessary *red shift* demand of the following layers. The *time shift* helps to provide in numbers and time what is missing in intensity. Fig.27.1 below demonstrates the effect of the time shift.

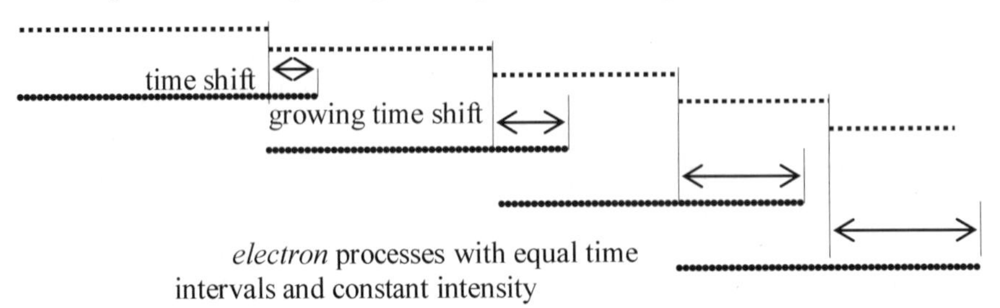

Fig. 27.1

Fig.27.1

Ref S.31

The effect of (1) the increased intensity of the proton and neutron processes and (2) the *time shift* create certain *blue shift* potential. Examples are presented in Section 31. This potential is a kind of *blue shift* deficit and *blue shift* provision towards the periphery. Each stabilised rotating status creates the necessary *blue shift* distribution – in fact with unchanged electron presence within the nucleuses of elements in rotation.

At the phase of accelerating up, the periphery of the disc is in high *blue shift* demand, which at constant peripheral speed is stabilised. At this stabilised status the periphery has a kind of saturated *blue shift* impact and potential – result of the *blue shift* surplus – relative to all other spots of the rotating disc, especially to the centre of the rotation with no *blue shift* conflict.

S. 27.2

27.2
Rotation with external electron flow

A rotating disc is taken and spots *X*, *Y*, and *Z* are marked on its radius in Fig.27.2.

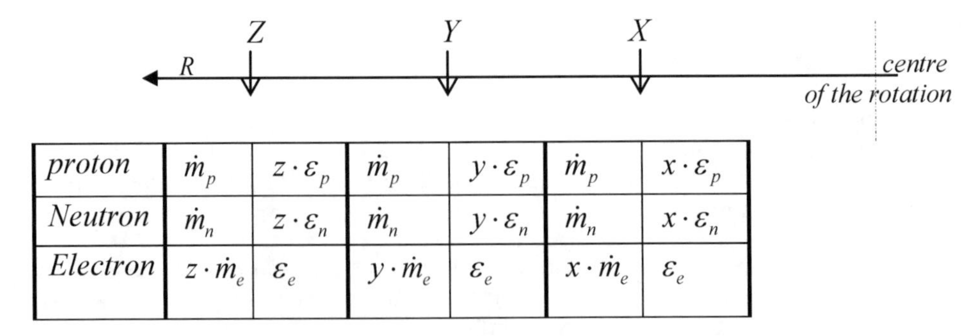

proton	$\dot{m}_p$	$z \cdot \varepsilon_p$	$\dot{m}_p$	$y \cdot \varepsilon_p$	$\dot{m}_p$	$x \cdot \varepsilon_p$
Neutron	$\dot{m}_n$	$z \cdot \varepsilon_n$	$\dot{m}_n$	$y \cdot \varepsilon_n$	$\dot{m}_n$	$x \cdot \varepsilon_n$
Electron	$z \cdot \dot{m}_e$	ε_e	$y \cdot \dot{m}_e$	ε_e	$x \cdot \dot{m}_e$	ε_e

Fig 27.2

Fig. 27.2

$\dot{m}_p$, $\dot{m}_n$ and $\dot{m}_e$ denote the mass change, ε_p, ε_n and ε_e are the intensities of the particles in normal conditions. *z, y*, and *x* in Fig.27.2 are multipliers, related to the spots at different radiuses, reflecting the increased values of the intensity.

With reference to 26F7 $\frac{1}{\varepsilon_R\sqrt{1-(v^2/c^2)}}\frac{mc^2}{dt_o}\left(1-\sqrt{1-\frac{(c-i)^2}{c^2}}\right);=\frac{\Phi_e}{\sqrt{1-(v^2/c^2)}}=\Phi$ Ref 26F7

the missing *blue shift* demand of the rotation, at radiuses, are:

$$\left.\begin{array}{l}\dot{e}_x=\Phi_x\cdot\\ \dot{e}_y=\Phi_y\cdot\\ \dot{e}_z=\Phi_z\cdot\end{array}\right.\dot{m}_e c^2\left(1-\sqrt{1-\frac{(c-i)^2}{c^2}}\right);\quad \text{and}\quad \Phi_z>\Phi_y>\Phi_x \qquad 27F1$$

and the relation of the intensity demand in *blue shift* energy is: $\dot{e}_z>\dot{e}_y>\dot{e}_x$

We can also calculate the frequency of this increased *blue shift* demand:

$$\Phi_x\frac{dE}{dt_i\varepsilon_x}=\frac{dn}{dt_i\varepsilon_e}q;\quad \Phi_x\frac{\dot{e}_x}{\varepsilon_x}=\frac{f_e}{\varepsilon_e}q;\quad \Phi_x\cdot\dot{e}_x\neq f_e\cdot q\quad \Phi_x\cdot\dot{e}_x=f_x\cdot q \qquad 27F2$$

$$f_{x,y,z}=\frac{\varepsilon_{x,y,z}}{\varepsilon_e}f_e$$ is the *blue shift* frequency demand of the neutron process

Φ_x,Φ_y and Φ_z in 27F2 denote the increased *red shift* demand of the neutron collapse. For covering the *blue shift* need, the *time shift* of electrons, acting within the disc is available, but in addition an external electron flow is connected to the rotating disc, as it is shown in Fig.27.3.

Is the *blue shift* effect of more electrons equivalent to the higher frequency of the *red shift* demand? No, but here the intensity of the *red shift* demand of the neutron process will be covered by the increased number, corresponding to the intensity demand.

The increased number of electrons will not just cover the *red shift* demand of the neutron process, but the *blue shift* surplus also will strengthen the *magnetic field* around the disc. The *time shift* and the electron flow increases the *blue shift* presence towards the periphery. The strength of the magnetic field will be corresponding to the *blue shift* surplus and will be increasing towards the periphery.

>> The magnetic field is a potential conflict to any similar magnetic fields or impacts of similar frequency;

There will be a certain resistance increase within the disc, consequence of the *blue shift* conflict caused by the *time shift* of the electrons of the disc and the additional electron flow.

>> The *blue shift* conflict of the flow is used as acting light *impact*.

Electrons, flowing towards the periphery and creating magnetic field can be used for another purpose: impacting the *Quantum Membrane*.

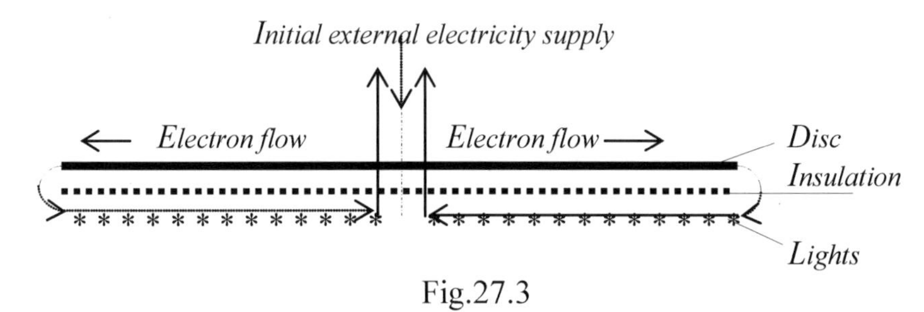

Fig 27.3

Fig.27.3

The amperage of the current, circulating within the two loops in Fig.27.3, corresponds to the maximal electron flow within the disc. This generates frequency f in each light spots of the rotating disc at stationary status.

27F3
$$\frac{dW}{dt_i\varepsilon_e} = \frac{dn}{dt_i\varepsilon_e}q$$

where W characterises the work of the conflict, generating the light impact; ε_e is the intensity of the electron process of the element of the disc at its stationary status, or in the centre of the rotation.

27F4
27F3 is equivalent to $\frac{dW}{dt_i} = \frac{dn}{dt_i}q$ or $\dot{w} = f \cdot q$

The rotation changes the intensity of the elementary processes. At certain speed n [1/sec] of the rotation of the disc, (measured within the system of reference of the *Earth*, taken as stationary), the intensities at spots *X, Y,* and *Z* of the radius will be of different values.

Ref 26F7

With reference to 26F7, the electron *blue shift* demand of the rotation is:

27F5
$$\Phi = \frac{\varepsilon_o}{\varepsilon_R\sqrt{1-(v^2/c^2)}} = \frac{\varepsilon_o}{\varepsilon_{Rv}} = \frac{\Phi_e}{\sqrt{1-(v^2/c^2)}}$$

27F6
and per the length of the radiuses: $\Phi_z > \Phi_y > \Phi_x$

The number of the quantum impact of the light effect is the same, but the frequency of the impact at different length of the radius, consequence of the intensity change, caused by acceleration, is different.

The longer the radius is, the more intensive is the need in electron *blue shift*. It also means that in the case of maximal external electron flow within the loops, with the growth of the radius, the frequency of the light impact is less and less. Increased electron flow at low intensity results in increased conflict. Therefore the light impact theoretically is increasing towards the centre of the disc with low intensity of the elementary (proton and neutron) process. The more intensive is the need, the less intensive is the conflicting impact: light.

The light impact is decreasing with the increase of the speed of the rotation. (This is only valid, if the number and the quality of light spots alongside the circumferences at different radiuses is the same. Otherwise the unequal number influences the frequency share.)

The quantum impact at the centre, or at stationary status is: $\frac{dW}{dt_i \varepsilon_e} = \frac{dn}{dt_i \varepsilon_e} q$ 27F7

The same quantum impact at rotation, at different radiuses: $\frac{dW}{dt_i \varepsilon_{z,y,x}} = \frac{dn}{dt_i \varepsilon_{z,y,x}} q$ 27F8

The same impact has different effects at different intensities (acceleration):

$$\frac{\dot{w}}{\varepsilon_e} \neq \frac{\dot{w}}{\varepsilon_{z,y,x}} \quad \text{and} \quad \frac{f}{\varepsilon_e} q \neq \frac{f}{\varepsilon_{z,y,x}} q \qquad \text{and if only } f \cdot q = f_{z,y,x} \cdot q$$

$$f_{z,y,x} = \frac{\varepsilon_e}{\varepsilon_{z,y,x}} f$$

since $\varepsilon_z > \varepsilon_y > \varepsilon_x > \varepsilon_e$ as consequence: $f_z < f_y < f_x < f$ 27F9

27F9 means: at any constant energy flow within the cycle of the rotating disc and with equal number of light spots at the circumference of different radiuses, the frequency of the light impact is *decreasing* towards the periphery.

(There is no contradiction with 27F2! 27F2 characterises the frequency of the *blue shift* demand. 27F9 is about the impact (light generation) of the electron flow. Since the amperage of the current within the cycle is constant, the frequency of the light impact at the spots towards the periphery is decreasing.) This effect can however been impacted by the number of the light equipment, which number towards the periphery will certainly be increasing and with that the utilisation of the electron flow as well.

The energy potential of the periphery of the rotating disc is different with external electricity supply than without it.

With reference to Section 27.1, the internal *blue shift* surplus of the rotation generates potential difference between the periphery and the centre. There is no potential difference between the spots of the disc without external electricity supply at stationary status. The value of the current corresponds to the maximal *blue shift* conflict between the *time shift* effect of the internal electron process and the *blue shift* of the external electron flow.

The energy potential of the periphery of the rotating disc is different with external electricity supply than without it.

S. 27.3

27.3
Benefit: lifting and fly

Rotating disc creates potential difference – electron flow – within its elementary structure. Electron flow creates light *blue shift* conflict and impacts the *Quantum Membrane*. In the case of a free rotating disc in parallel with the surface of the *Earth*, the *red shift* of the *blue shifted* impact creates lifting effect.

The *blue-shift* at collision with the surface of the *Earth* is:

27G1
$$\Delta e_{blue} = e\left(1-\sqrt{1-\frac{(g\Delta t)^2}{c^2}}\right)$$

Δt is the time while the acceleration of the surface of the *Earth* would reach the spot of the impact: characteristic of the energy impact of the collision with the *Earth*. More explanation on this time is given in Sections 9 or 23.

The value of the *blue shifted* impact at the surface of the *Earth* is:

27G2
$$e_{blue} = e\left(2-\sqrt{1-\frac{(g\Delta t)^2}{c^2}}\right)$$

The acting *benefit* – the *red shift* at the surface of the rotating disc – at level l of the original impact is:

27G3
$$\Delta e_{red} = e\left(2-\sqrt{1-\frac{(g\Delta t)^2}{c^2}}\right)\left(1-\sqrt{1-\frac{(g\Delta t)^2}{c^2}}\right)$$

The measured (already *red shifted*) impact of the detection at the disc – at the same level of the original impact is:

27G4
$$e_{red} = \Delta f_R\sqrt{1-\frac{(g\Delta t)^2}{c^2}} = e\left(2-\sqrt{1-\frac{(g\Delta t)^2}{c^2}}\right)\sqrt{1-\frac{(g\Delta t)^2}{c^2}}$$

With reference to Section 9 and 3C3 the expression of the height of the disc above the surface of the *Earth* is:

$$l = \frac{c^2}{g}\left(1-\sqrt{1-\frac{(g\Delta t)^2}{c^2}}\right) \approx \frac{g(\Delta t)^2}{2}; \qquad 27G5$$

In the case of the *red shift* at a lower level – lower than the original impact from the rotating disc – the benefit in 27G3 will be less, but the increased by the rotation original impact can be kept constant. The impact toward the *Earth* can be repeated and the *blue shift* reflection of the *Earth* multiplied.
With reference to 15J1, the benefit (the *red shift*) at the lowered spot is:

$$\Delta e_{lower-red} = e\left(2-\sqrt{1-\frac{(g\Delta t)^2}{c^2}}\right)\left(1-\sqrt{1-\frac{(g\Delta t_h)^2}{c^2}}\right) \qquad 27G6$$

and the measured impact at lowered levels is:

$$e_{lower-red} = e\left(2-\sqrt{1-\frac{(g\Delta t)^2}{c^2}}\right)\sqrt{1-\frac{(g\Delta t_h)^2}{c^2}} = e \qquad 27G7$$

where Δt_h belongs to the lower height above the *Earth* (since $\Delta t_h < \Delta t$)

27G7 gives: $$\sqrt{1-\frac{(g\Delta t_h)^2}{c^2}} = \frac{1}{2-\sqrt{1-\frac{(g\Delta t)^2}{c^2}}} \qquad 27G8$$

With reference to 15J2 on the relation of the height and the time, the modified height of the surface of the disc toward the *Earth* is:

$$l_h = \frac{c^2}{g}\left(1-\frac{1}{2-\sqrt{1-\frac{g^2\Delta t^2}{c^2}}}\right) \qquad 27G9$$

The infinite sequence of *blue-red shifts* at the surface of the *Earth* and at the surface of the rotating disc as well as the infinite sequence of conflicts of the *blue shifted* frequency and the magnetic field of the rotating disc – as shown in Fig.27.4 – may create a lifting thrust. In the case of lifting effect, with reference to 27G9, the surface of the disc towards the surface of the *Earth* is changing.

Fig 27.4

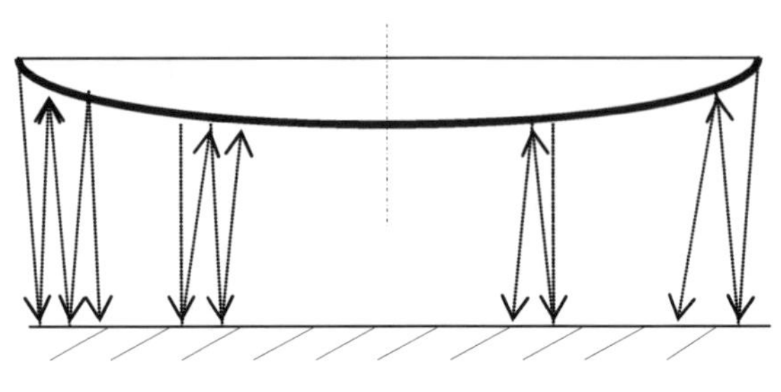

Fig. 27.4

The energy of the sphere symmetrical expanding acceleration of the *Earth* will be utilised in the following way:

- The light impact of the electron process on the rotating disc to the *Quantum Membrane* is *blue shifted* at the surface of the *Earth*;
- The *blue shifted* frequency will be partially in conflict with the magnetic field of the disc, creating lifting effect;
- The *red shift* at modified height, with reference 27G3, results in the original impact;
- The *blue shift* conflict with the magnetic field and the *blue/red shift* process starts with the same impact again, just from a lower height;
- The process will be repeated until the parabolic surface can provide a reasonable necessary height difference.

Fig 27.5 /a, /b

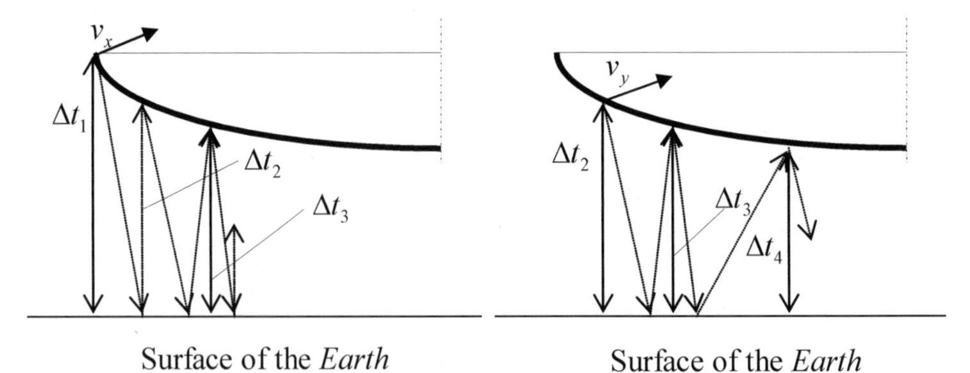

Surface of the *Earth* Surface of the *Earth*

Fig.27.5/a Fig.27.5/b

The summarised energy intensity benefit of the rotating parabolic disc with impact of a spot with peripheral speed v_x and radial acceleration of a_x – with multiplied *blue* and *red shift* sequence – is:

$$\sum e_x = e\left[\left(1-\sqrt{1-\frac{g^2\Delta t_1^2}{c^2}}\right)+\left(1-\sqrt{1-\frac{g^2\Delta t_2^2}{c^2}}\right)+\ldots+\left(1-\sqrt{1-\frac{g^2\Delta t_{n-1}^2}{c^2}}\right)\right] \quad \text{27H1}$$

Indexes of Δt in 27H1 relate to different, decreasing height values towards the centre of the disc, above the surface of the *Earth*, function of the parabolic surface. The *blue/red shift* sequence is shown in Fig.27.5/a.

The summarised impact of the sequence starting at another spot with speed v_y, closer to the centre of the rotation, as shown in Fig.27.5, is:

$$\sum e_y = e\left[\left(1-\sqrt{1-\frac{g^2\Delta t_2^2}{c^2}}\right)+\left(1-\sqrt{1-\frac{g^2\Delta t_3^2}{c^2}}\right)+\ldots+\left(1-\sqrt{1-\frac{g^2\Delta t_{n-1}^2}{c^2}}\right)\right] \quad \text{27H2}$$

And for any other peripheral speed of v_k or v_n:

$$\sum e_k^v = e\left[\left(1-\sqrt{1-\frac{g^2\Delta t_k^2}{c^2}}\right)+\left(1-\sqrt{1-\frac{g^2\Delta t_{k+1}^2}{c^2}}\right)+\ldots+\left(1-\sqrt{1-\frac{g^2\Delta t_{n-1}^2}{c^2}}\right)\right] \quad \text{27H3}$$

The peripheral speed means that all spots of the whole circumference with the same radius have the same peripheral speed.

There are n impacting spots on the parabolic surface with multiplied effect, the consequence of the *blue-shift/red-shift* sequence at the surface of the *Earth* and the parabolic surface.

The *blue/red-shift* sequence transfers the energy intensity of the sphere symmetrical expanding acceleration of the *Earth* to the rotating disc with parabolic surface.

$$\sum e = \sum e_x^v + \sum e_y^v + \ldots + \sum e_k^v + \ldots + \sum e_n^v \quad \text{27H4}$$

The result is either generated electron flow within the solenoid coils of the disc with parabolic surface or lifting magnetic thrust above the surface of the *Earth*, as shown in Fig.27.6.

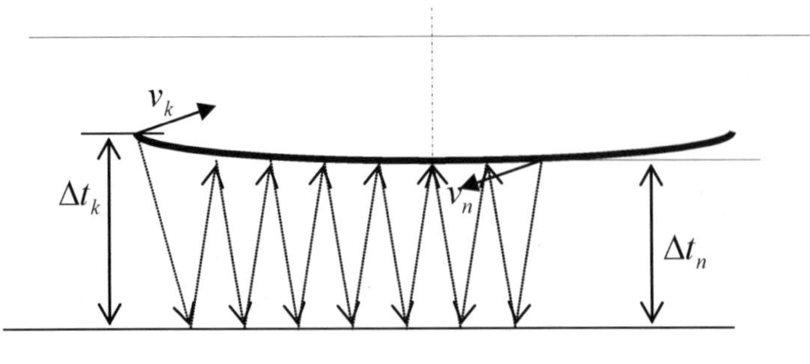

Fig
27.6

Fig.27.6

(The results and conclusions of the experiments with rotating discs with and without external electricity flow are given in Section 31.)

28

S. 28

Intensity reserve: *Energy Quantum*

There are two forms of acceleration in the elementary world: one, which is impacting the *Quantum Membrane* and the other, which is not.

There is no impact during the *proton process*:

$$\Delta e_p = \frac{dmc^2}{dt_p\varepsilon_p}\left(1-\sqrt{1-\frac{v^2}{c^2}}\right) = \frac{dmc^2}{dt_p\varepsilon_p} - \frac{dmc^2}{dt_p\varepsilon_p}\sqrt{1-\frac{v^2}{c^2}}; \quad \lim v > 0 \text{ and } v = i \qquad \text{28A1}$$

ε_p is the intensity of the speed increase – acceleration

The process in 28A1 can be understood in two ways:

- the mass values of the transformation into energy are equal, motion however speeds up the time flow and at speed *v*, dm, the mass in transformation (acceleration) needs more time; or
- the mass values of the change for equal dt_p time periods – with the growth of the speed – are decreasing.

These two explanations exist together and 28A1 means: during acceleration, mass looses from its "mass" capabilities: it transforms into energy.

(There is no alternative explanation, since $\frac{dmc^2}{dt_o\varepsilon_o} - \frac{dmc^2}{dt_v\varepsilon_v} = 0$ means matter is constant.) 28A2

The time system during the sphere symmetrical expanding acceleration (proton process) is changing. It is speeding up from $\lim v = 0$ to $i = \lim a\Delta t = c$. No impact to the *Quantum Membrane*.

There is a *Quantum Membrane* impact during the *electron process*:

$$\Delta e_e = \frac{dmc^2}{dt_i\varepsilon_e}\left(1-\sqrt{1-\frac{(c-i)^2}{c^2}}\right) = \frac{dmc^2}{dt_i\varepsilon_e} - \frac{dmc^2}{dt_i\varepsilon_e}\sqrt{1-\frac{(c-i)^2}{c^2}}; \qquad \text{28B1}$$

ε_e is the intensity of the $(c-i)$ acceleration of the electron process

The time system of the electron process is constant. Mass looses from its value, but not because of its transformation, rather for the work against the *Quantum System of Reference*: loading the *Quantum Membrane*.

With reference to 24C4, the *Quantum Membrane* can be loaded if

Ref 24C4

$$\mu c^2 - \mu c^2 \sqrt{1 - \frac{(c-i)^2}{c^2}} > q$$

Ref 24C3 28B2

at the same time, with reference to 24C3, the value of *quantum entropy* must be:

$$\dot{m}_{qe} = \mu \sqrt{1 - \frac{(c-i)^2}{c^2}} \geq q; \qquad \text{where } \dot{\mu} = \frac{dm}{dt_i dn}$$

The value of the *quantum entropy*, relative to a single quantum is:

28B3

$$\frac{dmc^2}{dndt_i}\sqrt{1 - \frac{(c-i)^2}{c^2}} \geq \frac{q}{dt_i}; \text{ or } \frac{dmc^2}{dndt_o}\sqrt{1 - \frac{i^2}{c^2}}\sqrt{1 - \frac{(c-i)^2}{c^2}} \geq \frac{q}{dt_i}$$

The *neutron process* is sphere symmetrical accelerating collapse – result of *red shift* – impact from the *Quantum Membrane*:

28C1

$$\Delta e_{ni} = \frac{dm_{qe}c^2}{dt_i \varepsilon_n} - \frac{dm_{qe}c^2}{dt_i \varepsilon_n \sqrt{1 - \frac{(i-v)^2}{c^2}}};$$

ε_n is the intensity (acceleration) of the collapse

The neutron process only happens under *Quantum Membrane* impact. All time and mass change consequences are result of this impact.
The meaning of 28C1 is:

- with the increase of the speed of the collapse, the time count of the mass system of reference is slowing down, therefore the collapse of equal mass values need less and less time; or
- during equal time periods of the process, the value of the collapsing mass (transformation of energy into mass) is increasing.

$u = (i - v)$ is the speed difference relative to *i*, the basis. At $u = 0$ speed difference, the absolute speed of the sphere symmetrical collapse is $v = i$

28C3

$$dt_i = \frac{dt_o}{\sqrt{1 - (i^2/c^2)}}$$

It is mathematically correct, but there is no such status, as $v = 0$, at rest.

Therefore at full collapse, it also can be written:

$$dt_i = \frac{dt_{(u=\lim v=0)}}{\sqrt{1-\frac{[c-(\lim v=0)]^2}{c^2}}} = \frac{dt_{\lim v=0)}}{\sqrt{1-\frac{i^2}{c^2}}} \quad \text{and} \quad dt_{\lim v=0} = dt_i\sqrt{1-\frac{i^2}{c^2}} \qquad 28C4$$

The energy taken by the full collapse is:

$$\Delta e_n = \frac{dmc^2}{dt_o\varepsilon_n}\sqrt{1-\frac{i^2}{c^2}}\sqrt{1-\frac{(c-i)^2}{c^2}} - \frac{dmc^2}{dt_o\varepsilon_n\sqrt{1-\frac{i^2}{c^2}}}\sqrt{1-\frac{i^2}{c^2}}\sqrt{1-\frac{(c-i)^2}{c^2}} =$$

$$= \frac{dmc^2}{dt_o\varepsilon_n}\sqrt{1-\frac{i^2}{c^2}}\sqrt{1-\frac{(c-i)^2}{c^2}} - \frac{dmc^2}{dt_o\varepsilon_n}\sqrt{1-\frac{(c-i)^2}{c^2}}; \qquad 28C5$$

At full collapse, with reference to 28C3 $dt_o = dt_p = dt_n$ 28C6

For comparison of the energy balance of the proton-electron-neutron cycle, the result in 28C5 shall be related to the proton system of reference.

The balance between the proton (*sphere symmetrical expanding acceleration*) and neutron (*sphere symmetrical accelerating collapse*) processes – in ideal case – is:

$$\frac{dmc^2}{dt_o\varepsilon_p}\left(1-\sqrt{1-\frac{i^2}{c^2}}\right) = \frac{dmc^2}{dt_o\varepsilon_n}\sqrt{1-\frac{(c-i)^2}{c^2}}\left(\sqrt{1-\frac{i^2}{c^2}}-1\right) \qquad 28C7$$

The intensity of the proton and neutron processes are taken in 28C7 as quasi equal:

$$\varepsilon_e = \frac{1}{Z} = \left|\frac{\Delta e_p}{\Delta e_n}\right| = \left|\frac{\varepsilon_p}{\varepsilon_n}\sqrt{1-\frac{(c-i)^2}{c^2}}\right| \approx 1$$

The distinguishing feature of elements is the difference in proton and neutron intensities.

28C7 can be written as:

$$\frac{\dot{m}_p c^2}{\varepsilon_p}\left(1-\sqrt{1-\frac{i^2}{c^2}}\right) = \frac{\dot{m}_n c^2}{\varepsilon_n}\sqrt{1-\frac{(c-i)^2}{c^2}}\left(\sqrt{1-\frac{i^2}{c^2}}-1\right) \qquad 28C8$$

and the relation of measured masses between *P* and *N* is $\frac{N}{P} = \frac{\dot{m}_n}{\dot{m}_p}\sqrt{1-\frac{(c-i)^2}{c^2}}$

Neutron and proton processes run fully in parallel and always in mass balance, but are shifted in time and are of different intensities. The electron process is the one, establishing the balance between them.

$\frac{dm}{dt_p} \neq \frac{dm}{dt_n}$; which is equivalent to $\dot{m}_p \neq \dot{m}_n$ but/and $\frac{\dot{m}_p}{\dot{m}_n} = \frac{\varepsilon_p}{\varepsilon_n}\sqrt{1-\frac{(c-i)^2}{c^2}} = \varepsilon_e$

Electrons are products of the proton process, and
- in the case of the $\varepsilon_p > \varepsilon_n$

there always will be an increasing electron *blue shift* surplus (free electrons) available.

- in the case of the $\varepsilon_n > \varepsilon_p$

  there will be an electron *blue shift* deficit, with two consequences: the *blue shift* available will always be fully used – there will not be any free electrons – and the neutron process will always be limited by the *blue shift* available within the element. Therefore external *blue shift* will be "welcome" and used even for the normal neutron process, since it naturally needs it.

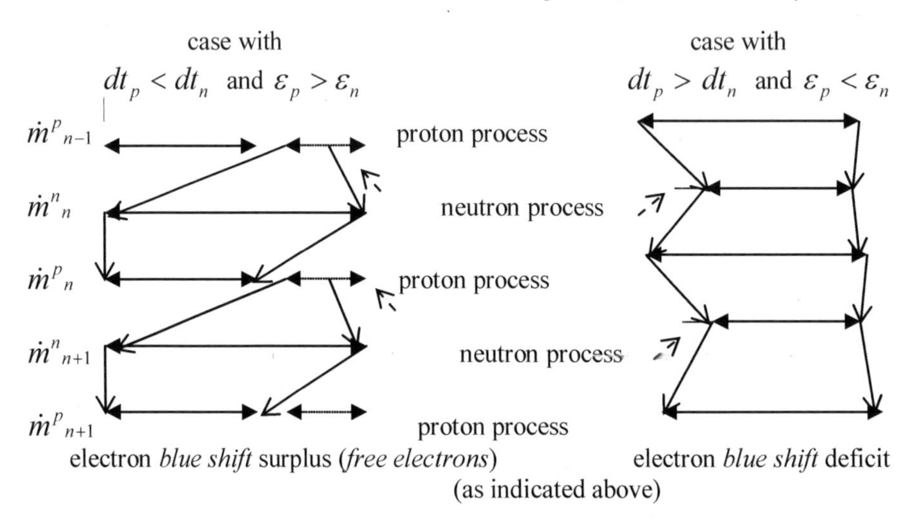

Diag. 28.1

Diag.28.1

For better explanation, $\dot{m}^n{}_n = \dot{m}^p{}_n = \dot{m}$ taken, and $\dfrac{dm^p}{dt\varepsilon_p} = \dfrac{dm^n}{dt\varepsilon_n} \rightarrow \dfrac{\dot{m}^p}{\varepsilon_p} = \dfrac{\dot{m}^n}{\varepsilon_n}$

if $\varepsilon_p > \varepsilon_n \rightarrow \dot{m}_p > \dot{m}_n$ - each cycle produces *free electrons* (with *blue shift* surplus)

if $\varepsilon_n > \varepsilon_p \rightarrow \dot{m}_n > \dot{m}_p$ - neutron process of each cycle goes with less capacity than it should be (*blue shift* deficit)

Proton-neutron processes go in full balance.

At the time moment $\dfrac{dm}{dt} = 0$, (meaning: the growing to infinity $\dfrac{dm}{dt} < 0$ is not the case any more), the internal "energy" of the collapsing mass can withstand the pressure of the *Quantum Membrane*. The collapse turns into expanding acceleration, where $\dfrac{dm}{dt} > 0$.

At the time moment $\dfrac{dm}{dt} = 0$ (meaning: the decreasing to zero $\dfrac{dm}{dt} > 0$ is not the case any more) the expanding acceleration, from *quantum entropy* status turns into collapse.
The cycle process continuous with decreasing, continuously less and less mass values.

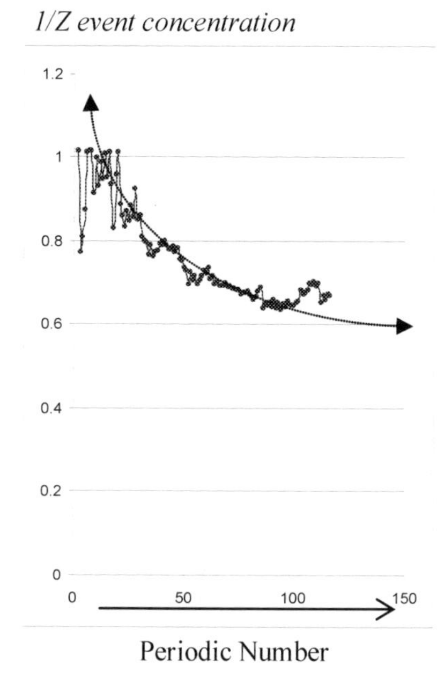

Diag.28.2/1

With the growth of the periodic number the *blue shift* deficit grows and neutron process looks for all possible *blue shift* impact to be utilised

Quantum System of Reference

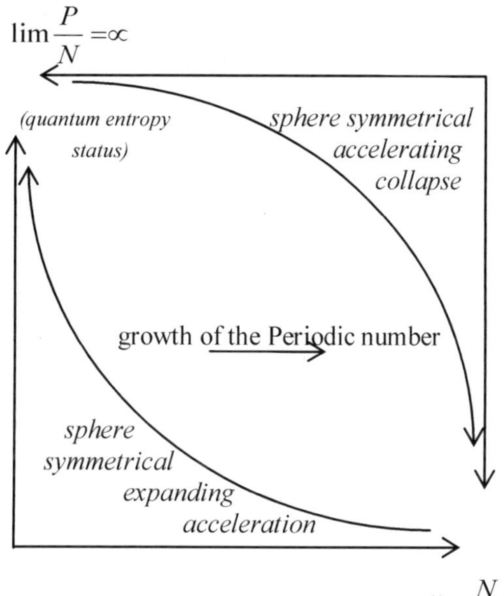

Unique feature of the mass status: no "escape" for *blue shift* impacts, all taken by neutron/s.

Diag.28.2/2

Diag. 28.2/1

Diag. 28.2/2

Intensity values of elementary processes *(dm/dt)* are changing during the process, since the time frame of the change is changing = mass is accelerating either in the direction of the expansion or the collapse. The intensity of the proton process is slowing down during the process (since the time frame is lengthening), the neutron process intensity – in the contrary – is growing (since the time frame is shortening). The change of the intensities follows the same hyperbolic function – but with different gradient values, which are the characteristic of elements.

Proton intensive elements are of light weight category, neutron intensive elements are the heavy ones. The more is the intensity of the neutron collapse, the heavier is the element.
Proton intensive elements are the active ones, stepping into energy balance relation ("chemical reactions") with others.

The interesting point is the start of the cycles.

$$\frac{dm}{dt} = \infty$$

The original infinite intensity value of the mass change accelerates mass to quantum status, and ends in infinite number of *quantum entropy*.

Matter is without change. Mass is transformed into energy and re-transformed back into mass. The balance however follows the *entropy law* with the benefit of the non-returning back into mass energy intensity component of the mass change. The value of the re-transformation for a single quantum is:

28D1
$$\Delta e = \frac{d\mu c^2}{dt_o}\sqrt{1-\frac{(c-i)^2}{c^2}}\left(\sqrt{1-\frac{i^2}{c^2}}-1\right); \text{ and}$$

$$e_q = \frac{d\mu c^2}{dt_o}\sqrt{1-\frac{(c-i)^2}{c^2}}; \quad \text{or} \quad q = \mu c^2\sqrt{1-\frac{(c-i)^2}{c^2}}$$ – is the *energy quantum* = the *non-re-transformed energy*, equivalent to the *quantum entropy*, reference to 28B2.

With the infinite cycle of the proton-electron-neutron process ongoing, the intensity of the proton mass change, after cycle n is:

28D2
$$\frac{dm_{(n+1)}}{dt_o} = \frac{dm}{dt_o}\left[\sqrt{1-\frac{(c-i)^2}{c^2}}\right]^{n+1}\left(\sqrt{1-\frac{i^2}{c^2}}-1\right);$$

and the accumulated quantum energy intensity reserve at the end of cycle n (the difference between the accumulated intensities of the proton and neutron processes) is:

28D3
$$\sum_0^n e_{qe} = \frac{dmc^2}{dt_o}\left[\sqrt{1-\frac{(c-i)^2}{c^2}}\right]^{n}\left(1-\sqrt{1-\frac{(c-i)^2}{c^2}}\right);$$

The *Quantum Membrane* acts in single quantum form. The proof is the response to the electron processes, which is ending with single *quantum entropy*. Should we explain 28D1 and 28D2, as "constant" *dm* mass change, related to changing time frame, the finding is that *matter* itself is accelerating – since the time count is growing.

S. 28.1

28.1 Acceleration from external source

Acceleration of protons and neutrons from external source means, the mass change is under two impacts in parallel:

- one is the internal expanding acceleration or the accelerating collapse,
- the other is the external acceleration of the internal "particle"-processes.

The description of the proton process, in motion with u, result of external acceleration is:

$$\Delta e_p^{external} = \frac{dmc^2}{dt_{po}} - \frac{dmc^2}{dt_{po}\sqrt{1-\frac{u^2}{c^2}}}\sqrt{1-\frac{v^2}{c^2}} = \frac{dmc^2}{dt_{po}} - \frac{dmc^2}{dt_{pv}\sqrt{1-\frac{u^2}{c^2}}} \quad \text{28E1}$$

For simplicity, the internal, element specific intensity value is not addressed in 28E1. (Conclusion on the intensity effect of the acceleration will be made on the assessed effect of the speed increase of the motion.)

$$dt_{pv} = \frac{dt_{po}}{\sqrt{1-\frac{v^2}{c^2}}} \quad \text{at } v=i \quad dt_i = \frac{dt_{po}}{\sqrt{1-\frac{i^2}{c^2}}}$$

result of the sphere symmetrical expanding acceleration (*SSEA*) 28E2

If $dt_{po}=1$ and $\varepsilon_{po}=1$ are taken, at speed v, result of the internal mass change of the *SSEA* (for the count of the mass-"energy" of the proton process) the intensity of the process is:

$$\varepsilon_{pv} = \frac{1}{dt_{pv}} = \frac{1}{dt_{po}}\sqrt{1-\frac{v^2}{c^2}} \quad \text{28E3}$$

In normal circumstances, without external acceleration – driven by the *internal* "energy" of the mass only – the intensity of the mass change is getting less and less, the duration of the process becomes longer.

The case is different with *external* energy.

- with reference to 28E1, external acceleration increases the intensity of the mass transformation (energy generation) of the proton process.

  Reference shall be given in the equation to the system of reference, relative to which the external acceleration is taking place.

  It seems difficult, but from the point of view of the system of reference of the proton in external acceleration, it can be specified: The proton is speeding up by speed u relative to this "unknown" system of reference. The description of the time system of the proton (which has already been modified by *SSEA*) reflects this. Acceleration is not reciprocal.

The time flow of the proton in *SSEA* relative to the system of reference of the external acceleration shall correspond to speed value u, and is:

$$dt_{pv} = \frac{dt_o}{\sqrt{1-(u^2/c^2)}} \quad \text{28E4}$$

where dt_o characterises the unknown system of reference

From 28E4 follows that the intensity of the proton mass transformation in acceleration from external source is:

28E5
$$\frac{dmc^2}{dt_{pv}\sqrt{1-(u^2/c^2)}} = \frac{dmc^2}{dt_{po}} \frac{\sqrt{1-(v^2/c^2)}}{\sqrt{1-(u^2/c^2)}}$$

The *SSEA* of the internal, elementary process is characterised by the time relation to speed v, the external acceleration is characterised by the time relation to speed u. Both relations refer to the time system of the proton.

The intensity of the proton process as result of the impact of the external acceleration is:

28E6
$$\varepsilon_{pu} = \frac{1}{dt_{po}} \frac{\sqrt{1-(v^2/c^2)}}{\sqrt{1-(u^2/c^2)}}; \qquad \text{and} \qquad \varepsilon_{pu} > \varepsilon_{pv}$$

At speed level v of the normal *SSEA* the intensity of the mass change becomes less.

With reference to 28E6, at speed u, result of acceleration from *external* source the intensity of the mass transformation *grows*. At speed $u = i$ the intensity growth is of infinite value.

We always have to keep in mind, that this all about the *intensity* of the mass change. The mass portion in change cannot become more than its original value, but the intensity of the transformation at increased speed of $u = i$ can be higher indeed than the original (starting) intensity. At full expansion of the mass and still at speed of $u = i$ the intensity of the mass change would be equal to the original starting mass change value.

The impact of external acceleration on the neutron process is similar to the proton process. The difference is that the time system of the neutron process at the start of the process belongs to speed $i = \lim a\Delta t = c$.

The elementary sphere symmetrical accelerating collapse (*SSAC*) is driven by external impact. The *blue shift*, generated by the electron process is acting as *red shift* and initiates the collapse. The *red shift* impact is not incorporating. The collapse itself is decrease of the speed for the count of the increase of the mass. (Energy is transforming back into mass.)

The description of the neutron process with external acceleration is:

28F1
$$\Delta e_n^{external} = \frac{dm_{qe}c^2}{dt_i} - \frac{dm_{qe}c^2}{dt_u} = \frac{dm_{qe}c^2}{dt_i} - \frac{dm_{qe}c^2}{dt_i\sqrt{1-\frac{(i-v)^2}{c^2}}\sqrt{1-\frac{u^2}{c^2}}} =$$

$$= \frac{dmc^2}{dt_{no}\sqrt{1-\frac{u^2}{c^2}}}\sqrt{1-\frac{(c-i)^2}{c^2}}\left(\sqrt{1-\frac{i^2}{c^2}}-1\right) \quad \text{28F2}$$

$$\left(\sqrt{1-\frac{(i-v)^2}{c^2}}-1\right) \text{ or } \left(\sqrt{1-\frac{i^2}{c^2}}-1\right) \quad \text{28F3}$$

- relate to the elementary *SSAC*
- at full collapse ($v = i$), the speed different is *i*.

Once external acceleration increases the intensity of the proton process, its direct consequence is the increase of the intensity of the neutron process as well. This is in line with the *continuity law* of matter:
- generation of energy shall be in balance with the use of energy;
- transformation of mass into energy shall be in balance with the re-transformation of energy into mass.

External acceleration similarly to the proton process, increases the intensity of the neutron process:

The intensity with the *SSAC* grows:
It is taken that at the start of the neutron process $dt_i = 1$; and $\varepsilon_i = 1$.

$$\varepsilon_{(i-v)} = \frac{1}{dt_i\sqrt{1-\frac{(i-v)^2}{c^2}}} \quad \text{28F4}$$

The external accelerating impact, similarly to the proton process, shall be expressed through the parameters of the system of reference of the acceleration. The relation of the time flow of the neutron process, at speed *u*, to the time flow of the system of reference of the acceleration impact is:

$$dt_{acceleraion} = dt_v\sqrt{1-\frac{u^2}{c^2}} \quad \text{28F5}$$

With reference to 28F1, the mass change relative to the system of reference of the external accelerating impact is:

$$\frac{dm_{qe}c^2}{dt_{acceleration}} = \frac{dm_{qe}c^2}{dt_i\sqrt{1-\frac{(i-v)^2}{c^2}}\sqrt{1-\frac{u^2}{c^2}}} \quad \text{28F6}$$

The intensity of the neutron process, impacted by external acceleration is growing:

28F7 $$\varepsilon_{nu} = \frac{1}{dt_i\sqrt{1-\frac{(i-v)^2}{c^2}}\sqrt{1-\frac{u^2}{c^2}}} = \frac{\varepsilon_{(i-v)}}{\sqrt{1-\frac{u^2}{c^2}}}; \quad \text{and} \quad \varepsilon_{nu} > \varepsilon_{(i-v)}$$

At $\lim u = i$ the intensity increase is of infinite value.

S 28.2

28.2
Accumulated intensity reserve

Proton and neutron transformations go with the change of the time system.

Blue shift however is an impact without the change of the time system. The process goes with constant time at speed $i = \lim a\Delta t = c$, acceleration for infinite time. The mass of the electron is changing: the *blue shift* loads the *Quantum Membrane*.

Red shift is impact of the *Quantum Membrane,* the drive of the neutron collapse with the mass of the neutron increasing. Its time system shortening and its intensity increasing.

Why the neutron process cannot happen without *red shift* impact?

The answer is coming from the status of *quantum entropy*, the final stage of the electron process.

The remaining mass "energy" intensity of the *quantum entropy* (as *blue shift* impact) cannot load any further the *Quantum Membrane*. The loaded *Quantum Membrane* takes over and under its "pressure" the *quantum entropy* collapse starts. Once the collapse started, the *red shift* of the *Quantum Membrane* permanently impacts the collapsing mass. This is not about the actual absolute mass values of the collapsing neutron, rather the permanent increase of dm/dt, the intensity of the collapse. The permanent increase deems external energy impact.

Quotient of dm/dt is growing and it cannot go without external (work) impact. The collapse stops, when the *Quantum Membrane* is not capable any more to increase the dm/dt quotient of the collapse.

This is why *red shift* impact cannot be incorporated: this impact is the drive of the collapse. What incorporating is "the energy", which is in fact

the change of the time system,

consequence of the proton process acceleration.

The time system, modified by the speeding up effect of the (internal) proton process is consolidating within the neutron process. The *Quantum Membrane* is the energy potential of the matter, accumulating as result of each proton-electron-neutron cycle.

Each proton-electron-neutron cycle produces – unresolved mass-time imbalance, hanging all over within the matter, accumulating as *Quantum System of Reference*, at the end of cycle n with reference to 28D3, value of:

$$\sum_0^n e_{qe} = \frac{dmc^2}{dt_o}\left[\sqrt{1-\frac{(c-i)^2}{c^2}}\right]^n\left(1-\sqrt{1-\frac{(c-i)^2}{c^2}}\right) \quad \text{Ref 28D3}$$

The new proton process always starts with:

$$\frac{dmc^2}{dt}\left[\sqrt{1-\frac{(c-i)^2}{c^2}}\right] \quad \text{28G1}$$

The proton- neutron difference is always:

$$\frac{dmc^2}{dt}\left(1-\sqrt{1-\frac{(c-i)^2}{c^2}}\right) \quad \text{28G2}$$

Infinity as such exists, therefore sphere symmetrical expanding acceleration of mass can last for infinite time. The intensity of the mass change however can never be *zero*. It would mean the stop of *time*, since without event – change – time does not exist.

(As there is no meaning of a formula with *zero* in the denominator as well.)

The general principle of entropy works: the imbalance in the cycle is part of the natural process: mass cannot fully transform itself into energy status. Mass acceleration reaches the status, equal to the value of *quantum entropy* and is not capable any more to impact the *Quantum Membrane*.

Each consecutive mass transformation cycle results in returning less (energy) intensity into mass – energy reserve is accumulating.

The quantum balance is

$$\mu c^2\left(1-\sqrt{1-\frac{(c-i)^2}{c^2}}\right) = qe$$

But the load only works, if

$$\mu c^2 - \mu c^2\sqrt{1-\frac{(c-i)^2}{c^2}} \geq qe$$

The *Quantum System of Reference* is in fact the accumulated intensity reserve of the mass transformation. *Time*, "product" of transformation of mass, result of the sphere symmetrical expanding acceleration of mass, cannot be fully *re-used* during the neutron process, the sphere symmetrical accelerating collapse, the re-transformation of energy into *mass*.

= This remaining intensity potential is the *energy quantum*.

Intensity, the gradient of the mass change reaches infinite low value at the end of the electron process of each cycle.

28G3 *Intensity* cannot be of *zero* value, since: $\varepsilon_i = \frac{1}{dt_i} = \frac{1}{(dt_o)}\sqrt{1-\frac{(\lim c)^2}{c^2}}$

But there is a *non-re-incorporating* intensity value, an *intensity reserve* unit – *energy quantum* – of the *Quantum System of Reference*.

The case can be demonstrated the following way:

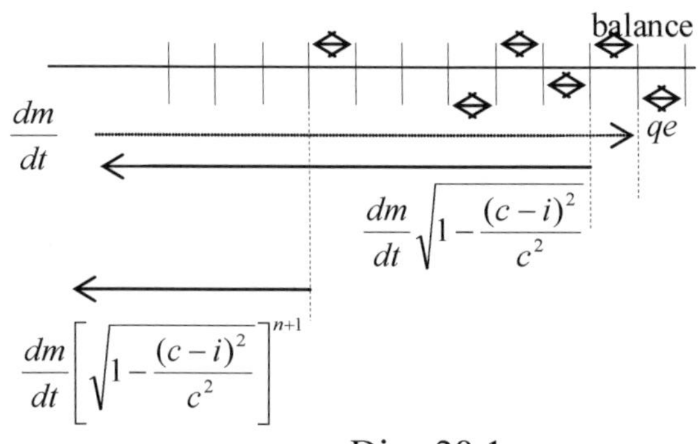

Diag 28.1

Diag.28.1

The *Quantum System of Reference* can be impacted and loaded. The impact is motion without time consequence (*blue shift* at constant speed). The response of the *Quantum Membrane* corresponds to the initial impact and either result in neutron *red shift* or in other kind of "external" impact.

S. 28.3

28.3
Intensity reserve = *Energy Quantum*

Proton transformation has negative mass change gradient:

28I1 $\frac{dmc^2}{dt_p}\sqrt{1-\frac{v^2}{c^2}} = -p$ with the increase of v the p is getting less and less

Neutron transformation has positive mass change gradient:

$$\frac{dmc^2}{dt_o\sqrt{1-\frac{(i-v)^2}{c^2}}}\sqrt{1-\frac{i^2}{c^2}}\sqrt{1-\frac{(c-i)^2}{c^2}}=n \qquad 28I2$$

with the increase of $u=i-v$, where $i \geq v > 0$ n is getting more and more.

The proton process is transformation of mass into energy and the intensity of the mass change is getting less and less. The neutron process is only possible with external energy intake, since the intensity of the mass change is increasing. The impact from the *Quantum Membrane* is the drive.

The proton and neutron processes directly relate to the time flow, but the decrease and increase of intensities during the mass change are in balance between the two processes. There is no impact to the *Quantum Membrane*.

The *Quantum System of Reference* is the accumulating intensity reserve of continuous proton-electron-neutron cycles.

The accumulated – and not utilised by the sphere symmetrical accelerating collapse of the neutron – intensity reserve, with reference to 28D3, is:

$$\sum_0^n e_{qe}=\frac{dmc^2}{dt_o}\left[\sqrt{1-\frac{(c-i)^2}{c^2}}\right]^n\left(1-\sqrt{1-\frac{(c-i)^2}{c^2}}\right) \qquad \text{Ref } 28D3$$

The time frame of the electron process is constant.
In this case, what the mass transformation of

$$\frac{dmc^2}{dt_o}\sqrt{1-\frac{i^2}{c^2}}\left(1-\sqrt{1-\frac{(c-i)^2}{c^2}}\right) \quad \text{is used for?} \qquad 28I3$$

The only target is the *Quantum System of Reference* to be impacted.

The accumulated intensity reserve of the mass transformation is impacted!
In other words, the *Quantum Membrane* is loaded.
Because of the constancy of the time system at $i=\lim a\Delta t=c$, the impact does not generate new intensity reserve (time quantum).

The impact cannot be a global one, increasing the level of the intensity reserve in "average", because all components of the *Quantum System of Reference* have been born as the smallest possible intensities – *energy quantum*.

28J1 The proton process starts with intensity: $\frac{dmc^2}{dt_o}$

28J2 and the neutron process ends with $\frac{dmc^2}{dt_o}\sqrt{1-\frac{(c-i)^2}{c^2}}$.

The difference is the intensity reserve of the mass change, the intensity which has not been re-transformed to mass:

28J3
$$\frac{dmc^2}{dt_o}\left(1-\sqrt{1-\frac{(c-i)^2}{c^2}}\right)$$

The mass value at the end of the neutron process is result of the sphere symmetrical accelerating collapse of the mass from the value of the quantum entropy:

28J4
$$\frac{dmc^2}{dt_o\sqrt{1-\frac{i^2}{c^2}}}\sqrt{1-\frac{i^2}{c^2}}\sqrt{1-\frac{(c-i)^2}{c^2}}=\frac{dmc^2}{dt_o}\sqrt{1-\frac{(c-i)^2}{c^2}}$$

The *quantum entropy* is developing in single form:

28J5
$$w_{blue-shift}=\frac{dmc^2}{dt_o}\sqrt{1-\frac{i^2}{c^2}}\left(1-\sqrt{1-\frac{(c-i)^2}{c^2}}\right)\geq q_{qe} \text{ and } q_{qe}=\frac{dmc^2}{dt_o}\sqrt{1-\frac{i^2}{c^2}}\sqrt{1-\frac{(c-i)^2}{c^2}}$$

Mass is collapsing, but the intensity reserve, equal to the summarised quantum entropy value stays as related to single processes.

The *Quantum System of Reference* shall be understood as:

28K1
$$\frac{dmc^2}{dt_o}-x\frac{dmc^2}{dt_o}\sqrt{1-\frac{i^2}{c^2}}\sqrt{1-\frac{(c-i)^2}{c^2}}$$

It can never mean:

28K2
$$\frac{dmc^2}{dt_o}-\frac{d(xm)c^2}{dt_o}\sqrt{1-\frac{i^2}{c^2}}\sqrt{1-\frac{(c-i)^2}{c^2}}; \quad \text{since } x\frac{dm}{dt_o}\neq\frac{d(xm)}{dt_o}$$

28K2 gives proof not just in mathematical, but also in physically terms: The intensity of the mass change cannot be more at the end of the sphere symmetrical accelerating collapse than it was at the beginning of the sphere symmetrical expanding acceleration.

Each quantum entropy develops its neutron process separately:

$$x\frac{dmc^2}{dt_o\sqrt{1-\frac{i^2}{c^2}}}\sqrt{1-\frac{i^2}{c^2}}\sqrt{1-\frac{(c-i)^2}{c^2}}=x\frac{dmc^2}{dt_o}\sqrt{1-\frac{(c-i)^2}{c^2}} \qquad \text{28K3}$$

28K3 means, whatever is the absolute value of the mass, the intensity corresponds to the intensity of the end stage of the neutron process.

S.
29

29

Speeding up results in *quantum communication*

The only process which can load the *Quantum Membrane* is the electron process, since the electron *blue shift* does not change the time system.
Infinite variety of the electron processes is equal to infinite variety of *blue shift* intensities, equal to certain number of – not used and accumulated – *quantum entropy* intensities.

29A1

$$\frac{dmc^2}{dt_i\varepsilon_x}\left(1-\sqrt{1-\frac{(c-i)^2}{c^2}}\right)=\frac{dn}{dt_i\varepsilon_x}q$$

ε_x in 29A1 is a specific intensity value of the electron process; it is specific to a certain element. It relates to the intensity of the mass change at constant time system of dt_i.

Ref
18D4
20A2
29A2

The intensity relation between the proton-electron-neutron processes is:

$$\varepsilon_e=\left|\frac{\varepsilon_p}{\varepsilon_n}\right|\sqrt{1-\frac{(c-i)^2}{c^2}}$$

Element specific intensity relations in general are about the values of the acceleration of the sphere symmetrical expansion and the collapse of the mass. The *continuity law of matter* between the proton and neutron processes corresponds to specific acceleration (intensity) values. This is the reason having elements with *blue shift* surplus and deficit.

Fig.29.1 demonstrates, proton and neutron processes follow the same function of change, just with opposite sign.

The intensities of the proton and neutron processes are symmetrical in general (as Fig.29.1 shows), but the intensity difference shifts the two processes in time (as Fig.29.2 shows.)

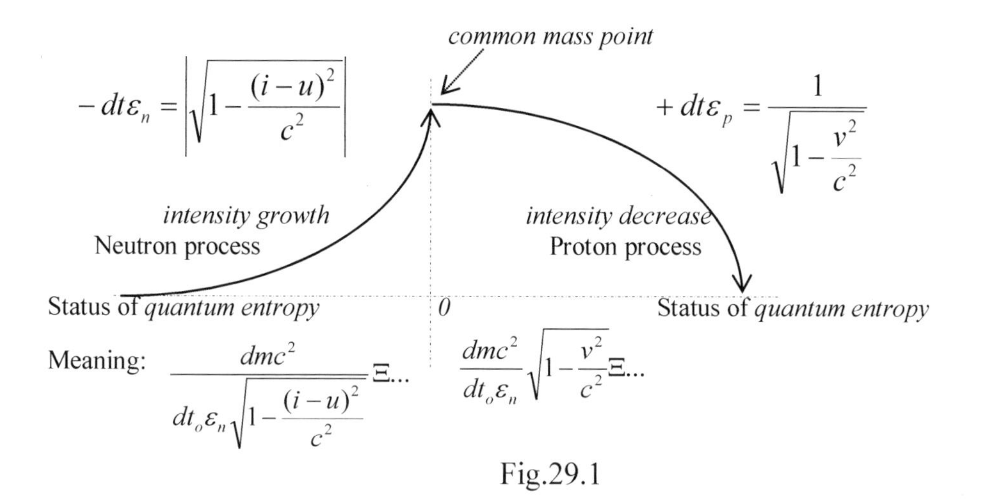

Fig.29.1

Fig 29.1

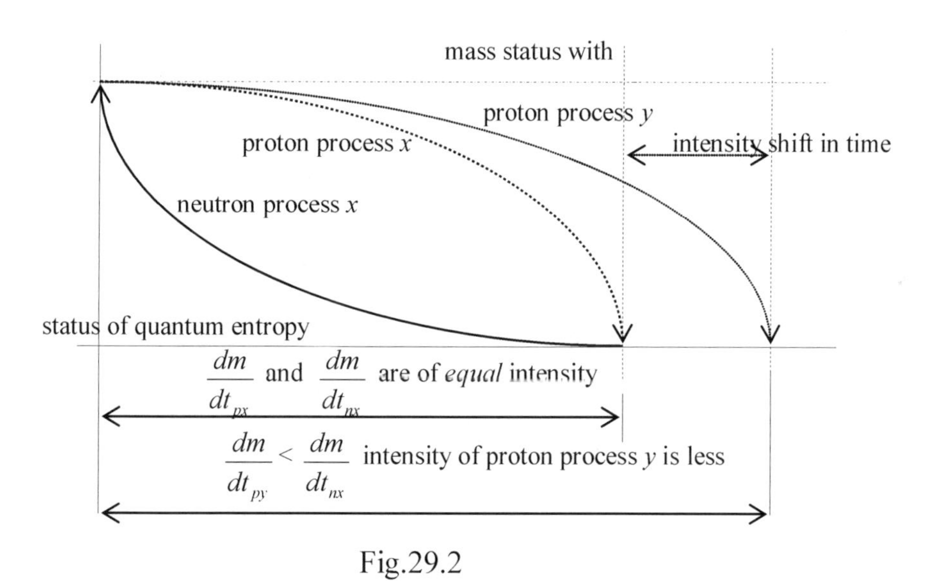

Fig.29.2

Fig 29.2

The general principle of the continuity law of matter is valid, but the intensities of reaching certain speed values during sphere symmetrical expansion and collapse depend on element specific data.

External intensity impact increases the intensity of both, the proton and neutron processes. The neutron and proton processes within elements, are harmonised in line with their intensity relations through the element specific intensity of the electron process: $\varepsilon_x = \varepsilon_i \cdot \varepsilon_e$.

Neutron collapse and proton expansion processes go in balanced way together. The continuity law of matter keeps them always in balance and combined. The time shift and the difference in their intensities are regulated by the electron process. The intensity relation with reference to 29A2, is always constant and guarantied.

Ref 29A2

External acceleration to the process impacts both, the neutron and the proton processes simultaneously. Should the neutron process be impacted and accelerated, the proton process will be accelerated and impacted the same way. External acceleration means harmonised increase of the intensities of both processes.

29B1 29B2 29B3

Neutron process	Proton process	Electron intensity is constant, without change
$\frac{mc^2}{dt_o\varepsilon_n}\frac{\sqrt{1-\frac{i^2}{c^2}}\sqrt{1-\frac{(c-i)^2}{c^2}}}{\sqrt{1-\frac{(i-u)^2}{c^2}}\sqrt{1-\frac{\upsilon^2}{c^2}}}$	$\frac{mc^2}{dt_o\varepsilon_p}\frac{\sqrt{1-\frac{v^2}{c^2}}}{\sqrt{1-\frac{\upsilon^2}{c^2}}}$	$\varepsilon_e=\left\lvert\frac{\varepsilon_p}{\varepsilon_n}\right\rvert\sqrt{1-\frac{(c-i)^2}{c^2}}$

- ε_p and ε_n, the proton and neutron intensities are regulated by the constant intensity of the electron process;
- external acceleration modifies (slows down) the time system of the neutron and the proton processes, increasing with that their process intensity, without modifying the original electron process intensity value of the element:

29B4 29B5

$$dt_{na}=dt_o\sqrt{1-(\upsilon^2/c^2)} \quad \text{and} \quad dt_{pa}=dt_o\sqrt{1-(\upsilon^2/c^2)}$$

The electron process is without change, controlling the intensified proton and neutron processes of the element.

S 29.1

29.1
Electron *blue shift* impact between elements

Reactions between elements are based on the *blue shift* impact of those, with *blue shift* surplus. These “active” elements with proton process dominance (in conventional terms with *free electrons*) initiate “chemical” reaction.
The *blue shift* is

$$\frac{dmc^2}{dt_i\varepsilon_x}\left(1-\sqrt{1-\frac{(c-i)^2}{c^2}}\right)$$

The *blue shift* of an "active" element is in fact an external impact – impacting (accelerating) the neutron process of other elements. With reference to 29B1 and 29B2, it results in certain increase of the intensity of the sphere symmetrical expanding acceleration and accelerating collapse.

Ref
29B1
29B2

The increased intensity of the neutron process results in balanced intensity increase of the proton process. From the point of view of the recipient element, the elementary process becomes intensified; from the point of view of the donor element the electron *blue shift* surplus is used. The balance is guarantied, since one of the elements is providing, the other is receiving the *blue shift*.

Elements vary in their distinguishing *time shift* and the intensity difference between the proton and neutron processes. Certain reaction of elements with *blue shift* surplus and deficit may result in extra stable, close to

$$\varepsilon_e = \frac{\varepsilon_p}{\varepsilon_n}\sqrt{1-\frac{(c-i)^2}{c^2}} \approx 1 \text{ elementary composition.}$$

Meaning: the integrated intensities of the proton and neutron processes of the compounds are quasi equal. Where it is possible, the *blue shift* surplus of certain elements can be used for covering the *blue shift* deficit or demand of others. Elements can create compounds and mixtures, where they utilise their *blue shift* "capabilities".

Electrons communicate this way, but external electron *blue shift* will never modify elements. The internal electron process of element is responsible for the proton/neutron relations.

29.2
External acceleration needs work and results in work effect – weight

S
29.2

As the experience shows, external (mechanical) acceleration of solid structures – with elements with dominant neutron process – needs intensive work load. External acceleration of elements in gaseous state – where the proton process is dominant – the work might be unnoticed in conventional circumstances.

The effect on the neutron process is obvious: it intensifies the transformation of energy into mass and results in increased gradient of mass change (in conventional terms – increased weight and mass status).

Ref 29B1 29C1

$$\frac{mc^2}{dt_n\varepsilon_n}\frac{\sqrt{1-\frac{i^2}{c^2}}\sqrt{1-\frac{(c-i)^2}{c^2}}}{\sqrt{1-\frac{(i-u)^2}{c^2}}\sqrt{1-\frac{\upsilon^2}{c^2}}}$$

The adjustment of the denominator by the increased speed clearly demonstrates the intensity growth: the time system is slowing down and the gradient of the mass change is growing

The effect of the acceleration on the proton process is similar, but the result is not mass increase rather the increase of the intensity of the mass transformation into energy:

Ref 29B2 29C2

$$\frac{dmc^2}{dt_p\varepsilon_p}\frac{\sqrt{1-\frac{v^2}{c^2}}}{\sqrt{1-\frac{\upsilon^2}{c^2}}}$$

The time system of the proton process is slowing down:

$$dt_p > dt_p\sqrt{1-\upsilon^2/c^2}$$

This can be interpreted in two ways: either the transformation of the same mass happens for less time or more mass transformation happens for the same time period

External acceleration has no direct effect on the electron process.

29C3

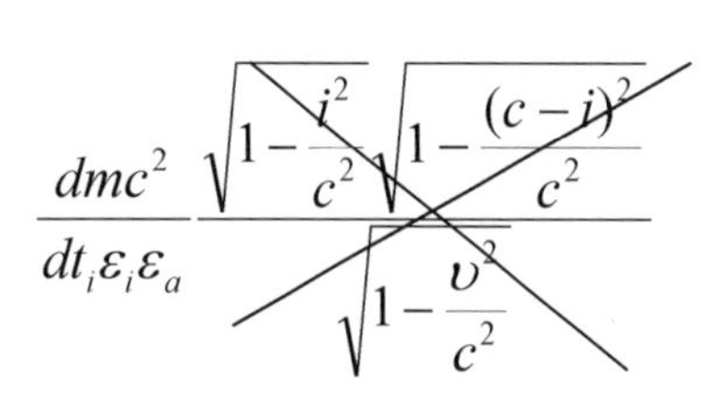

The $dt_i\sqrt{1-\upsilon^2/c^2}$ component within the denominator with υ different then zero makes no sense: the time system of the electron process cannot be different than dt_i and the intensity of the electron process for a certain element is constant

29C3 can only in physical terms is not just impossible but would result in double effect: the increase of the intensity of the neutron process is balanced by the increased proton change gradient. The increased intensity of the proton process results in *increased* number of *electron generation.* At the same time however the constancy of the intensity of the electron process has not been impacted.

External acceleration increases the mass transformation gradient of the *neutron* process. It directly means: external acceleration needs external work, proportional to the intensity increase of the neutron collapse – the increase of the mass change gradient. The *proton* process is a consequence and while the acceleration has a significant impact on the proton process as well, the proton process itself is not about mass gradient increase, rather "energy transformation gradient" increase.

In conventional terms this finding means: weight (in gravitational circumstances) is result of the increased intensity of the neutron process.

The question is where is the line when we measure the impact (weight) of the proton and neutron processes? The end of the neutron process and the start of the consecutive proton process are the same. When we measure the nucleus, we measure the effect of all particles together from the start of the neutron collapse to the full expansion of the proton and electron processes.

It also means that if work has been applied for acceleration, this work value proves and demonstrates the intensity increase of the neutron process. This is important to state because there is no way to consider proton acceleration without neutron acceleration. The overall balance of transformation and re-transformation does not allow accelerating "particles" separately.

29.3 External impact and speeded up time flow

S. 29.3

The impact to the *Quantum Membrane* is electron *blue shift*:

$$[Event] = \frac{dmc^2}{dt_{ii}}\left(1 - \sqrt{1 - \frac{(c-i)^2}{c^2}}\right) \quad \text{29D1}$$

The time relation of the electron *blue shift* of elements and the system of reference of the *Earth* is: $dt_{ii} = \dfrac{dt_{io}}{\sqrt{1 - \dfrac{i^2}{c^2}}}$

The event in 29D1 within the system of reference of elements is:

$$\frac{dmc^2}{dt_{ii}}[Event] \quad \text{29D2}$$

The same event, within the system of reference of the *Earth* is (measured as) of infinitely low intensity

$$\frac{dmc^2}{dt_{ii}}[Event] = \frac{dmc^2}{dt_{io}}\sqrt{1 - \frac{i^2}{c^2}}[Event] \quad \text{29D3}$$

Blue shift impact and/or conflict of this level of intensity as in 29D3 is usual and can be managed without any harm or left not to be noticed within the system of reference of the *Earth.* (Radio waves can be one of the many examples.)

For making the impact remarkable, the *number* of the generated electrons shall be increased. This is not about the modification of the intensities of the electron process!

The proton and neutron processes happen in different time system, with increased intensity. The electron process remains of the same intensity, but electrons are generated in increased number.

The mass change of the elementary process is in line with the internal balance. The description of the electron process with intensified electron generation is:

29D4 $$\frac{dmc^2}{dt_{io}\sqrt{1-\frac{v^2}{c^2}}}\sqrt{1-\frac{i^2}{c^2}}[Event]=\frac{dmc^2}{dt_{io}\sqrt{1-\frac{v^2}{c^2}}}\sqrt{1-\frac{i^2}{c^2}}\left(1-\sqrt{1-\frac{(c-i)^2}{c^2}}\right)$$

29D5 In the case of $v=i$ the process in 29D4 corresponds to the system of reference of the *Earth* and is of: $$\frac{dmc^2}{dt_{io}}[Event]$$

The intensity of the electron generation in 29D5 is infinite times more than the original one in 29D3.

If the increased in this way impact is loaded back to the original elementary system of reference, the effect within the elementary system will be further increased. It will be intensified infinite times again, as result of the external acceleration.

29D6 The process will be: $$\frac{dmc^2}{dt_{io}\sqrt{1-\frac{i^2}{c^2}}}[Event];$$ since $dt_{ii}=dt_{io}\sqrt{1-\frac{i^2}{c^2}}$

If we take a particle accelerator and accelerate elementary particles up to speed $i=\lim a\Delta t=c$, there will be two kinds of effects:

- the external acceleration will increase the intensity of the *proton-neutron* processes, resulting in increased number of electrons (processes);
- the increased number of electrons will result in increased electron process *blue shift* conflict with the accelerating *blue shift* effect of the accelerator.

  → Increased *blue shift* (and possible magnetic) conflict is source of energy: may generate heat.

At speed of $i=\lim a\Delta t=c$, the acceleration of particles cannot be separated from the *blue shift* impact of the sphere symmetrical expanding acceleration of the *Earth*. The *blue shift* impact of the *Earth* may result in further intensity growth (beyond the effect of the acceleration and the increased speed), resulting in additional *blue shift* and *magnetic* conflict.

29.3.1. Speeded up time flow

Quantum Membrane can "externally" be impacted by the increased number of electron processes. The increase in numbers results in ε_x modified intensity (in 29E1 below) and f_x modified frequency:

$$f_x \cdot q = \frac{dmc^2}{dt_o \varepsilon_e \varepsilon_x} \sqrt{1-\frac{i^2}{c^2}} \left(1-\sqrt{1-\frac{(c-i)^2}{c^2}}\right) = \frac{dn}{dt_o \varepsilon_e \varepsilon_x} q \rightarrow$$

electron process impact

$$\frac{dmc^2}{dt_o \varepsilon_n \varepsilon_x} \frac{\sqrt{1-\frac{i^2}{c^2}}}{\sqrt{1-\frac{(i-v)^2}{c^2}}} \sqrt{1-\frac{(c-i)^2}{c^2}} \rightarrow \frac{dmc^2}{dt_o \varepsilon_p \varepsilon_x} \sqrt{1-\frac{(c-i)^2}{c^2}} \left(1-\sqrt{1-\frac{i^2}{c^2}}\right) \rightarrow$$

end stage of the neutron collapse proton process of the next cycle

$$\rightarrow \frac{dmc^2}{dt_o \varepsilon_e \varepsilon_x} \sqrt{1-\frac{(c-i)^2}{c^2}} \sqrt{1-\frac{i^2}{c^2}} \left(1-\sqrt{1-\frac{(c-i)^2}{c^2}}\right) = \sqrt{1-\frac{(c-i)^2}{c^2}} f_x \cdot q \qquad \text{29E1}$$

electron process of the next cycle impact is decreasing

The increased number of the electron process → increases the *blue shift* impact to the *Quantum Membrane* → the impact will be transferred → elements with *blue shift* surplus might be in conflict, since they do not need external *blue shift* impact → those targeted elements, which need *blue shift* impact and are programmed for the receipt, will take it.

The transmission of the electron *blue shift* impact (as calculated frequency value) is not constant. As it approaches "wider and wider" *Quantum Membrane* "surfaces" – it looses from its value.

Infinite number of accelerated (increased in number) conflicting electron processes *may* impact the *Quantum Membrane*. The electron *blue shift* impact is:

$$\frac{dmc^2}{dt_i \varepsilon_x \varepsilon_{external}} \left(1-\sqrt{1-\frac{(c-i)^2}{c^2}}\right) = \frac{dmc^2}{dt_i \varepsilon_z} \left(1-\sqrt{1-\frac{(c-i)^2}{c^2}}\right) = \frac{dn}{dt_i \varepsilon_z} q \qquad \text{29E2}$$

the summarised intensity is $\varepsilon_z = \varepsilon_i \varepsilon_e \varepsilon_{external}$ and the impact is:

$$\frac{dmc^2}{dt_i \varepsilon_z} \left(1-\sqrt{1-\frac{(c-i)^2}{c^2}}\right) = \frac{dn}{dt_i \varepsilon_z} q = f_z \cdot q \qquad \text{29E3}$$

External impact increases the use of the *Quantum Membrane* and results in increased neutron process intensity at the end of the cycle:

29E4

$$\frac{dmc^2}{dt_o\varepsilon_{Xx}\sqrt{1-\frac{(i-v)^2}{c^2}}}\sqrt{1-\frac{i^2}{c^2}}\sqrt{1-\frac{(c-i)^2}{c^2}}=\frac{dmc^2}{dt_o\varepsilon_{Xx}}\sqrt{1-\frac{(c-i)^2}{c^2}}$$

The value of the denominator decreases, which leads to shortening of the time and increasing the intensity of the mass change. As consequence, external impact to the *Quantum Membrane* increases the *intensity* of the mass change, and *accelerates time in general.*

In the case of *n* simultaneous impacts to the *Quantum Membrane* it means

29E5

$$\frac{dmc^2[Event]}{dt_o\varepsilon_{Xx}\sqrt{1-\frac{(i-v)^2}{c^2}}}+\frac{dmc^2[Event]}{dt_o\varepsilon_{Xx}\sqrt{1-\frac{(i-v)^2}{c^2}}}+\ldots+\frac{dmc^2[Event]}{dt_o\varepsilon_{Xx}\sqrt{1-\frac{(i-v)^2}{c^2}}}=$$

$$=\frac{dmc^2[Event]}{\frac{1}{n}dt_o\varepsilon_{Xx}\sqrt{1-\frac{(i-v)^2}{c^2}}}$$

More intensive mass transformation means the transformation of the same mass, but for less time. Increased impact to the *Quantum Membrane*, result of increased external *blue shift* impacts and *blue shift* conflicts = *speeding up the time flow.*

S. 29.4

29.4 Hydrogen

The measured (by us) last element of the periodic table is the *Hydrogen*: proton mass in expanding acceleration, neutron mass in collapse with infinite low intensity for infinite time, electron process impacting the *Quantum Membrane*. How this happens?

All *Helium* neutrons end up as *Hydrogen* protons. The expanding acceleration of the proton of the *Hydrogen* can only happen, if the neutron collapse “makes the room” for it in the internal mass-energy balance of the *Hydrogen* element. Therefore the neutron collapse goes on, even with infinite long time shift and infinite low intensity. The stop of the neutron process would not just mean the stop of the *Hydrogen*, but also the stop of *time*. And this cannot happen!

With the ongoing *Hydrogen* neutron collapse – even with infinite slow tempo – the time system of the neutron process is changing: from *quantum entropy* status, where the time system belongs to $i = \lim c$, to $\lim v = 0$ status, where the intensity value of the change itself is close to infinity, with infinite demand of *red shift*. The proton mass is expanding and electrons are generated. The electron process reaches *quantum entropy* status and the neutron collapse starts again and again, collapsing with infinite low intensity for infinity…

In the case of other elements, different than the *Hydrogen*, at one point, the collapse cannot be increased any more and the expansion starts. The neutron collapse, at the stage of infinite intensity, when the intensity cannot be increased any further - turns into proton process. Room in the balance for the followed by the neutron process proton process in fact is made by the turn of the neutron itself.

The proton process (the expanding acceleration) can start with immediate effect. The collapse needs *blue shift* impact, but this *blue shift* impact exists at the time moment of the turn of the neutron to proton, since the general rule of entropy does not allow all *blue* as *red* shift used for the collapse. The quantum entropy ensures this way that the neutron collapse is always less than the proton acceleration.

The key of the elementary process is the neutron collapse. It determines the "room" for the proton process. The neutron collapse needs *red shift* and any kind of *blue shift* available can be used.

$$\frac{dmc^2}{dt_i\varepsilon_x}\left(1-\sqrt{1-\frac{(c-i)^2}{c^2}}\right)=\frac{dn}{dt_i\varepsilon_x}q=f_x\cdot q;\ \text{or}$$

$$\dot{m}_i c^2\left(1-\sqrt{1-\frac{(c-i)^2}{c^2}}\right)=\frac{dn}{dt_i\varepsilon_x}q=f_x\cdot q \qquad \text{29F1}$$

Neutron collapse does not need specific *blue shift* frequency. Elements with neutron *red shift* deficit can utilise available *blue shift* all around for their collapse. Elements are different and the difference is in their electron process intensity, function of the proton-neutron intensity relations.

Proton acceleration is: $\frac{dmc^2}{dt_p\varepsilon_p}\left(1-\sqrt{1-\frac{i^2}{c^2}}\right)$;

If the normal proton process is externally (for the count of an external energy source) accelerated, the acceleration of mass will be increased, which intensifies the expansion (even at the constancy of ε_p):

29F2
$$\frac{dmc^2}{dt_p \varepsilon_p \sqrt{1-\frac{u^2}{c^2}}}\left(1-\sqrt{1-\frac{i^2}{c^2}}\right) > \frac{dmc^2}{dt_p \varepsilon_p}\left(1-\sqrt{1-\frac{i^2}{c^2}}\right);$$

In the case of the neutron process, this external acceleration also means increased effect:

29F3
$$\frac{dmc^2}{dt_n \varepsilon_n \sqrt{1-\frac{(i-u)^2}{c^2}}}\sqrt{1-\frac{(c-i)^2}{c^2}}\left(1-\sqrt{1-\frac{i^2}{c^2}}\right)$$

The growth of speed difference $(i-u)$ intensifies the neutron process.

29F4 In the case of either $\frac{N}{P}>1$ or $\frac{N}{P}<1$ external acceleration and all other *blue shift* available will impact and intensify the neutron process, and with that the proton process as well.

In the case of the *Hydrogen* we have to be careful with external acceleration. The intensification of the – by the *nature* regulated infinite long – neutron process may result in uncontrolled *red shift* demand.

30

S. 30

Electron flow generation

Impact of elements in general means electron *blue shift* impact or *blue shift* conflict within the *Quantum Membrane*.

Earth is in sphere symmetrical expanding acceleration.

The *Pound-Rebka-Snider* experiment at *Harvard University*, with its *blue shift* results has proved – *Earth* is in expanding acceleration. The expanding acceleration of the *Earth* corresponds to electron process. (The proton process is transformation without impact to the *Quantum Membrane*, the neutron process is collapse.)

Earth, in motion with $i = \lim a\Delta t = c$ is impacting the *Quantum Membrane* and generates *blue shift*.

The intensity of the acceleration is: $\frac{dv}{dt} = \frac{(c-i)}{\Delta t} = g$ 30A1

The balance of the *blue shift* impact is:

$$\frac{dmc^2}{dt_{io}\varepsilon_E}\sqrt{1-\frac{i^2}{c^2}}\left(1-\sqrt{1-\frac{(c-i)^2}{c^2}}\right) = \frac{dn}{dt_{io}\varepsilon_E}q \qquad \text{30A2}$$

ε_E denotes the intensity, corresponding to the *gravitation* of the *Earth*.

The time system of the *Earth*, taken as
dt_{io} - relates to speed $i = \lim a\Delta t = c$, the sphere symmetrical expanding acceleration of the *Earth*.

The time system of the electron process of elements, taken as
dt_{ii} - also relates to $i = \lim a\Delta t = c$.

This similarity or adequacy in speed $i = \lim a\Delta t = c$ does not exclude the difference in the two time systems, since

30A3 $$dt_{ii} = \frac{dt_{io}}{\sqrt{1-\frac{i^2}{c^2}}}; \text{ or } dt_{iii} = \frac{dt_{ii}}{\sqrt{1-\frac{i^2}{c^2}}}; \quad \text{if we suppose that } dt_{i} = \frac{dt_{o}}{\sqrt{1-\frac{i^2}{c^2}}}$$

30A3 means, the measured duration of events within the elementary world are infinite long relative to their measured duration within the system of reference of the *Earth*. The elementary system may have similar time and speed relations with other systems of reference.

The difference in the time systems is the reason of the difference in the perception of events. The measured duration of the same event varies, depending on the time system of the system of reference of the measurement in motion.

If the relation of the time systems of the elementary world and the *Earth* is:

30A4 $$dt_{ii} = \frac{dt_{io}}{\sqrt{1-\frac{i^2}{c^2}}},$$

where dt_{io} - denotes the time system of the *Earth* and

dt_{ii} - the time system of the electron process of the elementary world.

the sensation of the event within the elementary system of reference is:	the same event, within the system of reference of the *Earth* will be percept as

$$\frac{dmc^2}{dt_{ii}}[Event] \neq \frac{dmc^2}{dt_{io}}[Event]$$

$$\frac{dmc^2}{dt_{ii}}[Event] \neq \frac{dmc^2}{dt_{ii}}\sqrt{1-\frac{i^2}{c^2}}[Event]$$

The *electron* process of the elementary system of reference, within the time system of reference of the *Earth* is:

30A5 $$\frac{dmc^2}{dt_{ii}}\left(1-\sqrt{1-\frac{(c-i)^2}{c^2}}\right) = \underline{\frac{dmc^2}{dt_{io}}\sqrt{1-\frac{i^2}{c^2}}\left(1-\sqrt{1-\frac{(c-i)^2}{c^2}}\right)}$$

perception (measurement) within the time system of the *Earth*

The equation above means: *the mass change (impact), measured within the system of reference of the Earth is infinitely week.*

This perception in the *Earth* system obviously might give the sensation of constancy rather than change, a stable status rather than process. But this shall not be misleading: the process is the same. The difference is *only* in perception.

The perception of the neutron and proton processes gives similar sensation.

The time system of the *Earth* and the time system of the elementary world are of different intensities. If we take that

$$dt_{ii} = \frac{dt_{io}}{\sqrt{1-\frac{i^2}{c^2}}}\text{; and the intensity of the } \textit{Earth} \text{ system is taken for } \varepsilon_{io} = 1, \qquad \text{30B1}$$

$$\text{the intensity of the elementary world is } \varepsilon_{ii} = \sqrt{1-i^2/c^2} \qquad \text{30B2}$$

– meaning: the intensity of the elementary world is infinite time less.

The two systems cannot "communicate" with this infinite difference in time systems and intensities.

If we want to bring the time systems of the *Earth* and the elementary world into communicating relation, we have to *externally speed up* the elementary process.

We take a neutron process and accelerate it to speed u:

$$\frac{dmc^2}{dt_{ii}\varepsilon_{in}\varepsilon_a}\frac{\sqrt{1-\frac{i^2}{c^2}}\sqrt{1-\frac{(c-i)^2}{c^2}}}{\sqrt{1-\frac{u^2}{c^2}}}\left(1-\frac{1}{\sqrt{1-\frac{(i-v)^2}{c^2}}}\right) \qquad \text{30B3}$$

if $u = i$ or close to i, the time system of the neutron will be equal to:

$$dt_{io} = dt_{ii}\sqrt{1-\frac{i^2}{c^2}}\text{ ; since } dt_{ii} = \frac{dt_{io}}{\sqrt{1-\frac{i^2}{c^2}}} \qquad \text{30B4}$$

as consequence, the two systems will be communicating.

30B3 can be written as:

$$\frac{dmc^2}{dt_{io}\varepsilon_{on}\varepsilon_a}\sqrt{1-\frac{i^2}{c^2}}\sqrt{1-\frac{(c-i)^2}{c^2}}\left(1-\frac{1}{\sqrt{1-\frac{(i-v)^2}{c^2}}}\right) \qquad \text{30B5}$$

ε_{on} - includes the intensity of the time system and the intensity of the element specific characteristics of the neutron process.

It is value of: $\varepsilon_{on} = \frac{\varepsilon_{in}}{\sqrt{1 - i^2/c^2}}$ increased in infinite time, corresponding to the intensity of the system of reference of the *Earth*.

ε_a - corresponds to the value of the external acceleration of the elementary process.

ε_a in 30B3 and 30B5 can be of any value, depending on the technical capabilities of the accelerating device. The acceleration (or intensity of the speed increase) will always relate to the actual speed value. If $\varepsilon_a = 0$ there is no acceleration and no speed increase, $v = 0$.

The external acceleration means:

- the neutron (and the proton) processes have been accelerated;
- the increase of the intensity of the *neutron* process, resulting in the *slowing down of the time system* of the neutron process: from dt_{ii} to dt_{io}. This is in fact the reason why the mass of the neutron in collapse is growing – the intensity of the process is growing (as the everyday practice of external acceleration demonstrates).

The time system of the neutron process will always be relating to the speed increase. Once the acceleration is *zero*, but the speed level is *v*, the time system will be corresponding to this *v* level. Without acceleration, the elementary process will correspond and return to its internal standard.

As result of the speeding up of the complete elementary process, the time system of the *Earth* and the time system of the elementary process can "communicate" and may have impact to each other on practical level.

The increased neutron process also means increased proton process intensity and increased generation of electrons.

Because the permanent *blue shift* impact of the *Quantum Membrane* works against the motion, the communication between the two time systems of reference (and their impact on each other) can only be maintained if the speed level is maintained.

Since the slowing down process mainly relates to the time system of the elementary process, for a certain long while even the slowing down will not be sensed (measured) in substance within the system of reference of the *Earth*.

External acceleration needs energy and in return generates electrons. This is a kind of electricity generation, but without additional benefit, since the acceleration needs the same energy. The real *benefit* is the utilisation of the *blue shift* impact of the *Earth*. It further intensifies the neutron collapse.

- The intensified neutron process results in intensified proton process and increased electron generation, equal to *free electron generation = electricity for use.*

The *blue shift* impact of the *Earth* -

$$\frac{dmc^2}{dt_{io}\varepsilon_E}\left(1-\sqrt{1-\frac{(c-i)^2}{c^2}}\right)= \qquad \text{30C1}$$

(within the elementary world means increased by $\frac{1}{\sqrt{1-i^2/c^2}}$ impact)

$$=\frac{dmc^2}{dt_{ii}\varepsilon_E\sqrt{1-\frac{i^2}{c^2}}}\left(1-\sqrt{1-\frac{(c-i)^2}{c^2}}\right)=\frac{1}{dt_{ii}\varepsilon_E}\frac{dn}{\sqrt{1-\frac{i^2}{c^2}}}q \qquad \text{30C2}$$

- RESULTS IN INCREASED *ELECTRON FLOW*

- The generated free electrons might be not taken away. This will cause *blue shift* conflict and intensive *heat generation.*

- If the generated *blue shift* corresponds to the *blue shift* value of *gravitation* – since the *blue shift* conflict is *magnetic conflict* – it results in magnetic thrust and mechanical leverage force between the *gravitation* of the *Earth* and the electron *blue shift* of the element in acceleration.

S. 31

31

Rotating Disc Experiments

The objective of the experiments was to demonstrate: acceleration intensifies elementary processes. Two experiments were made: one with electricity supply and the other, without. Equipment and tools were manufactured in the Timar Workshop in Paks, Hungary, measurements were made in the same workshop together with Gabor Timar and Istvan Balogh in 2009 and 2010.

It is well known that rotation creates potential difference and also that rotating disc has its certain effect on the electricity flow. The importance of the findings is in their *quantum explanation.*

S. 31.1

31.1 Rotating Disc – Generator of Electron Potential

The device is a construction of 2 homogenous *Aluminum* discs (diameter of 362 mm, thickness of 5.82 mm), installed on a common rotating axle. Discs are insulated from each other and from the axle of the rotation. The axle is rotated by an electromotor, with bearings on both sides of the disc.

First measurements were made only with a single disc. Second disc was installed between the drive and the working disc for shielding, as much as possible the magnetic field of the electromotor.

The potential difference between the periphery and the centre of the discs is measured through *Carbon* brush contacts. *Al* disc is one of the best for generating and demonstrating the potential.

The accelerating effect of the *rotation* generates potential difference between the periphery and the centre of the rotating *disc* without external electricity supply.

The potential difference means difference in electron *blue shift* impact alongside the growing radius. It is consequence of the *intensity* and the *time shift* impact of the acceleration, as it is described in Section 27. The longer is the radius and higher the speed of the rotation the higher is the potential difference.

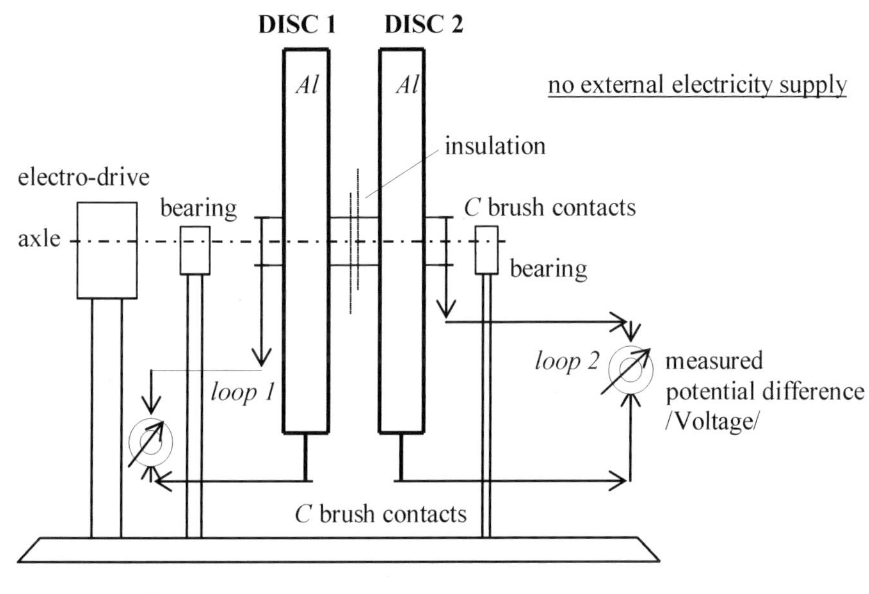

Fig. 31.1

Fig. 31.1

After the additional shield-disc was installed, the distance between the electromotor and the discs was increased with a longer axle.

Demonstration of the growing *blue shift* demand of the neutron process at the start of the rotation

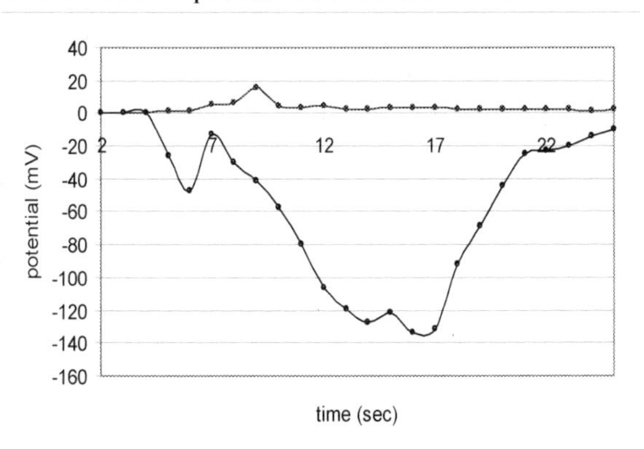

With reference to Section 27, the *acceleration* intensifies the proton and the neutron processes. The consequence is increased electron *blue shift* demand and deficit of the neutron process. Diag.31.1 shows this growing demand in the first 14 seconds of the rotation – result of the increasing speed. After reaching the 6000 RPM the *time shift* of the electron process works and the demand is covered.

Diagram 31.1.

Diagr. 31.1

The upper line is the potential within the shield-disc. The increasing *blue shift* demand of the periphery – result of the acceleration is covered by the *blue shift* of the magnetic field of the electromotor. This experiment was made with the two disc device, but the motor and the discs were at a short distance to each other. That is why the impact of the motor on the shield-disc is so significant.

The measured potential means electron flow from the centre to the periphery.

The experiment with single disc device has provided important information on the acceleration and the influence of the electromotor. There is no disc-shield in the following two examples. The magnetic field of the motor has its certain impact on the results.

The circumstances and the place of the measurements have influenced very much the results. The speeding up time from 0 to 6000 RPM is with growing *blue shift* demand. After it was consolidating.

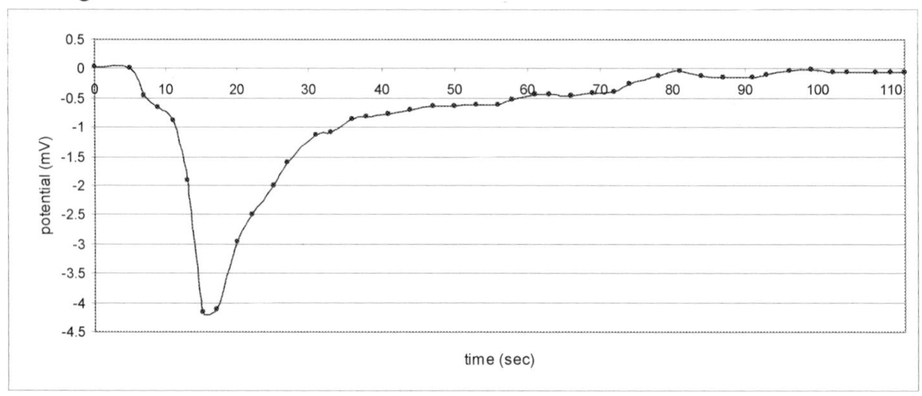

Diagr. 31.2

Diagram 31.2

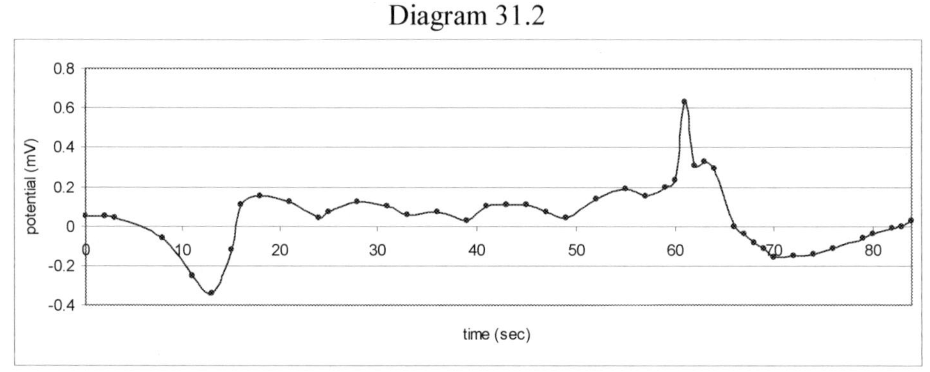

Diagr. 31.3

The current flowing through the measurement device was between (– 35 nA) and (+35nA)

Diagram 31.3

Diagram 31.3 shows the significance of the *time shift* of the electron process: at the moment of the slowdown – at the moment of the turning off the electromotor – the "accumulated" electron potential, caused by the *time shift* of the electron process within the disc, at the periphery reaches its maximum.

0-13 sec: is the speeding up: acceleration. The intensity growth of the proton and the neutron processes is dominant; there is an electron *blue shift* deficit in the direction of the growing radius. The potential of the periphery is negative relative to the centre.
13-18 sec: the electron process *time shift* gives the compensation (stable speed rotation at 6000 RPM – acceleration only in the direction of the radius).
Time period until 59th sec the status is very much stabilised.
At 59 sec, at the moment of the stop of the drive the accumulated reserve of the electron process *time shift* causes a maximum potential value.

The measurements in Diagram 31.4 were the latest ones. It was not just the distance between the motor and the discs increased from 450 mm to 900 mm, but the electromotor was covered with additional shielding metal net.

The results reflect the change: the *blue shift* impact of the magnetic field of the electromotor still has its effect, but it is less. The increasing intensity impact of the acceleration during speeding up is demonstrated in both discs. There is no compensation, or if any, they impact both disc the same way.

Electromotor is protected and drive distance from the discs increased.

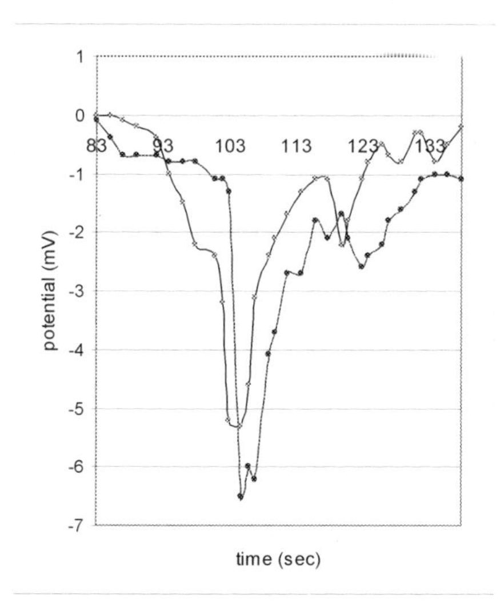

Diagr.31.5/a

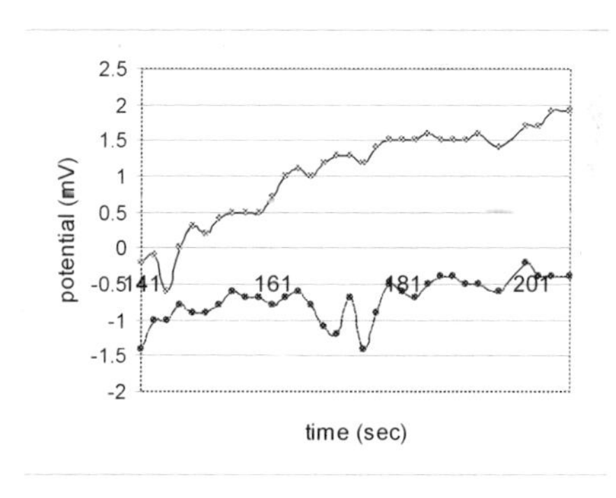

The first 4 diagrams belong to the same acceleration process and show the case: the deficit, the consolidation, the permanent speed period, the slow down and the stop

Diagr.31.4/b

Diagr. 31.4 /a, /b

Diagr. 31.4 /c, /d

Diagr. 31.4 /e, /f

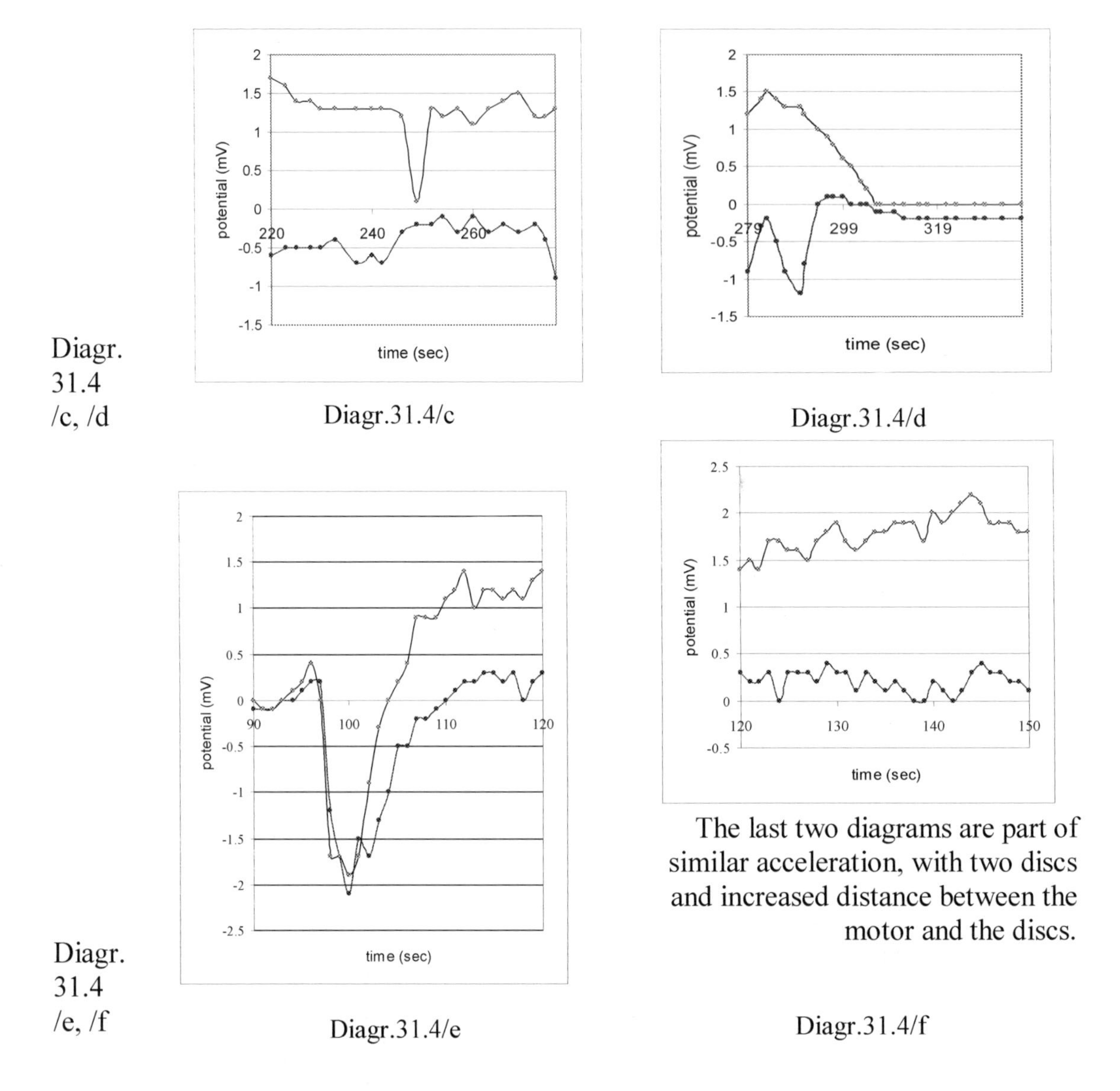

Diagr.31.4/c

Diagr.31.4/d

Diagr.31.4/e

Diagr.31.4/f

The last two diagrams are part of similar acceleration, with two discs and increased distance between the motor and the discs.

Before isolating the electromotor, pictures like in Diagram 31.5 with two discs also were measured. The *blue shift* deficit is simply missing.

Diagram 31.5 means, there is a *blue shift* transfer between the periphery and the centre within the "working disc"; the impact of the rotation on the disc-shield has been compensated by other *blue shift* sources, including electromotor.

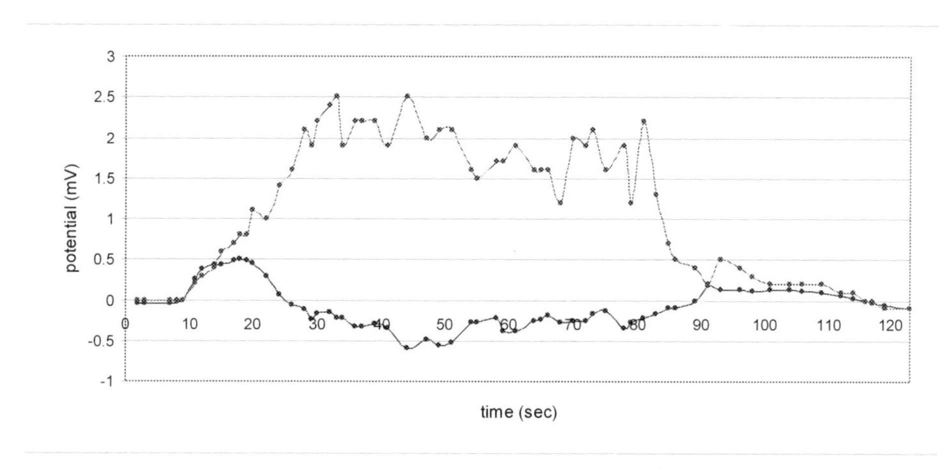

Diagram 31.5

Diagr. 31.5

In different way, but all results prove: the *acceleration* of the rotation increases the potential difference between the *periphery* and the *centre* of the rotation. The circumstances heavily influence the character, the development and the value of the difference.

The reason of this potential difference is the impact of the *acceleration* on elementary processes. The intensity of the mass-energy transformation, represented by the proton and neutron processes is directly impacted. The intensity of the electron process remains unchanged. It causes however electron deficit and surplus within the elementary process.

31.2. Rotating Disc – Generator of Quantum Impact

S. 31.2

The objective of the experiment remains the same, but the impact of the acceleration (of the rotation) is assessed with electron flow connected to the disc.

The elementary electron *blue shift* demand and deficit of the neutron process of the rotating disc – with external electricity supply – shows a specific feature: stimulating the external electron flow towards the periphery.

The device is the on Fig.31.2.

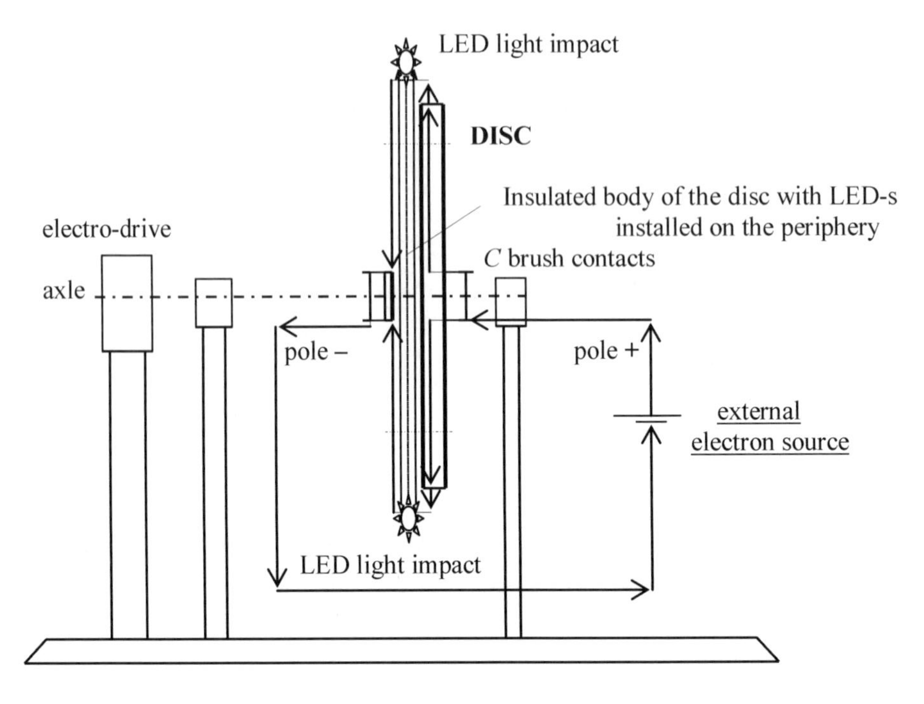

Fig. 31.2

Fig. 31.2

The *device* is a construction with a single homogenous working disc, installed on a rotating axle. *Light Emitting Diodes* (LED-s) are installed in parallel all around the periphery of an insulated spacer disc, fixed to the rotating disc. The positive pole of the external electron source is connected to the centre of the *"working"* disc through a *Carbon* brush contact. The negative pole of the external source is connected via similar *C* brush contact to a small support disc on the other side of the working disc. The insulation spacer disc is positioned between the working disc and the small supporting disc. The positive and negative poles of the LED-s are connected to the working disc and to the supporting small disc and with those to external supply respectively. The rotating axle is installed on bearing supports.

The cycle of the electron flow is the following: external electron source towards the positive pole of the centre of the working disc – electron flow within the working disc towards the periphery – electrons are approaching the positive poles of the *Light Emitting Diodes* – electron *blue shift* conflict causes light impact within the LED-s – electrons flow from the negative poles of the LED-s towards the negative "exit" C brush contacts at the centre of the small disc – and get back to the external source.

"Working" disc means the disc, internal elementary process of which fully determines the event and controls the process.

The diameter of the disc and the structure, the direction and the size of the support stand is indifferent. The only criterion is the free rotation of the disc. The material of the disc is indifferent, but the quantum impact of the LED is only measurable with discs of metal structures.

The accelerating effect of the *rotation* increases the intensity of the elementary (proton-neutron) process within the structure of the *working disc*. This is the reason, why the rotation stimulates electron flow in the direction of the growing radius. With the growing radius and with the growing speed the demand in electron *blue shift* grows.

The accelerating effect of the rotation generates *blue shift* conflict between the internal electron process of the disc and the external electron flow. The resistance of the disc against the external electron flow demonstrates the conflict. The light impact of LED-s corresponds to the value of the resistance. The higher is the resistance, the stronger is the light generation (impact to the quantum membrane).

In the case of external electron source with *unlimited* capacity: With the increase of the speed of the rotation – the *blue shift* conflict and the resistance grows – the unlimited external electron flow covers the need: the light impact of the LED-s is strengthening.
In the case of limited external electron source with *constant* potential (accumulator): With the increase of the speed of the rotation – the growing intensity increases the electron flow – at the same time the light impact of LED-s, because of the less *blue shift* conflict, weakens.

The *blue shift* demand of the rotation and the increasing electron flow from external electron source towards the periphery results in electron *blue shift* surplus.

For testing the resistance of the periphery of the rotating disc, a wire of permanent resistance was spooled on a rotating stainless steel disc, diameter of 250 mm. The change of the speed of the rotation from zero up to 6000 RPM and back resulted in changing electron flow within the wire under constant voltage of $U = 6.4$ V from an accumulator, which is in this case represents limited electron source. The findings are shown in Diagram 31.6.

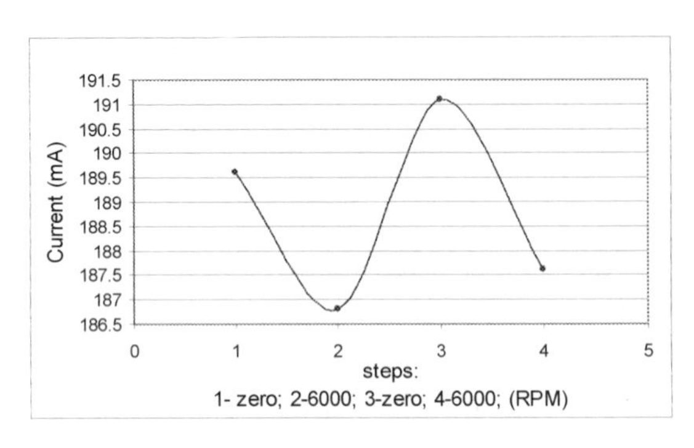

Diagr. 31.6

Diagram 31.6

Measurement of the current within the wire of permanent resistance at the periphery of the rotating disc (speed of the rotation is 6000 RPM and zero, U=6.4 V constant from accumulator)

The *blue shift* surplus at the periphery of the rotating disc results in increased resistance.

The measured current within the wire at speed 6000 RPM is less than at rest. The *blue shift* surplus at the periphery represents an increased *blue shift* conflict to any external electron flow. (We can interpret this effect in a different way as well: the increased *blue shift* demand at the periphery utilises the *blue shift* of the electrons flowing within the wire, therefore the measured electron flow between the two ends of the wire is less.)
With slowing down the resistance conditions of the current are "improving".

The increase of the resistance of the disc, as result of the rotation (even without external electron flow) can be directly measured. Diagram 31.7 shows the data: Results of the measurement of the resistance of two *Al* discs in rotation, connected in parallel <u>*without*</u> any external electricity supply.

Discs are 5.82 mm thick and the diameter of 362 mm. (Al-93.4%; Si-5.25%; Fe-0.8%; Mg-0.04%; Mn-0.04%.) *R* is measured between the *C* contacts on the axle and the *C* contacts on the periphery.

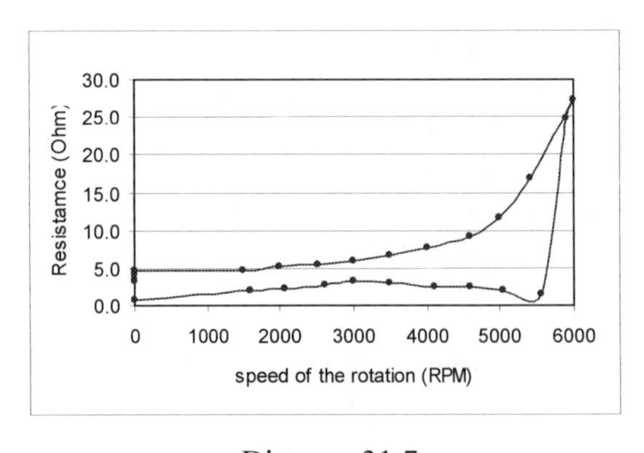

Diagr. 31.7

Diagram 31.7

The resistance was measured during the speeding up from 0 up to 6000 RPM (the bottom line) and during slowing down (the other line).

The diagram demonstrates the increased resistance of the periphery at 6000 RPM, result of the *blue shift* flow toward the periphery.

This resistance exists even if the *blue shift* surplus is created by the *time shift* of the electron process.

[*The increased mass impact (weigh or centrifugal force) of the rotating disc* towards the periphery *is result of the increased intensity of the neutron collapse* (re-transformation of energy in to mass), result of external work.]

If external current is connected to the disc with LED-s on the periphery, with one end connected to the centre and the other to the periphery, the potential difference between the two poles will be changing during the increase of the speed of the rotation. Measurements are presented on Diagram 31.8.

LED lights are on, the speed of the rotation is changing, the initial voltage, current, light and temperature are measured.

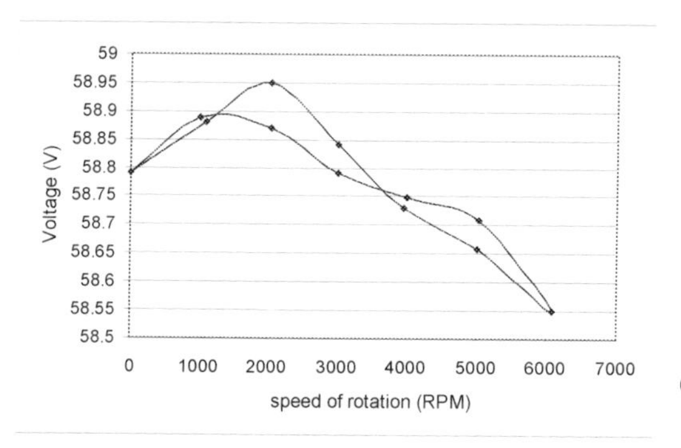

Diagram 31.8

The explanation of this phenomenon is that the increase of the intensity of the proton-neutron process, consequence of the speeding up results in increased *blue shift* demand.

The value of the current from external electron source follows the demand dictated by the rotation.

Diagr. 31.8

The value of the external electron flow corresponds to the *blue shift* demand of the periphery of the working disc, the difference between the *time shift* effect of the internal electron process of the working disc and the *blue shift* need of the rotation.

During the first period, the internal electron flow of the disc covers the *blue shift* need. This results in increased *blue shift* conflict at the periphery against external electron flow, which results in increasing potential difference.

With the further increase of the speed of the rotation, with reference to Diagram 31.8, at approximately at 2000 RPM of the rotation the external electron flow takes over and will be dominant. The intensified elementary process towards the periphery results in increased electron flow demand – the resistance obviously decreases, as the driving potential difference as well.

For good measure we have to note that the potential which can be created by the *time shift* of the internal electron process is of range of 2-3 mV. Here, the external potential difference in effect between the poles at the periphery and the centre is at 58 V level, which is quite a pressure to move electrons within the disc. The potential difference between the periphery and the centre of the disc is changing. At level of 2-3 mA of the electron flow, the difference is around 200 mV. The range of the change of the potential during the full length of the speed increase is 400 mV.

It also can be noted that the disc in rotation without external electricity supply itself "generates" around 3-4 nA within the wires of the measurement – as impact of the changing intensity status of the periphery and the centre of the disc.

The intensity increase, consequence of the rotation will be impacting the elementary process of the LED-s as well: As result of the increased neutron (and proton) intensity, the process goes with less *blue shift* conflict. The light impact of LED-s therefore at speed is less than at stationary status.
Experiments prove it and Diagram 31.9 shows the results.

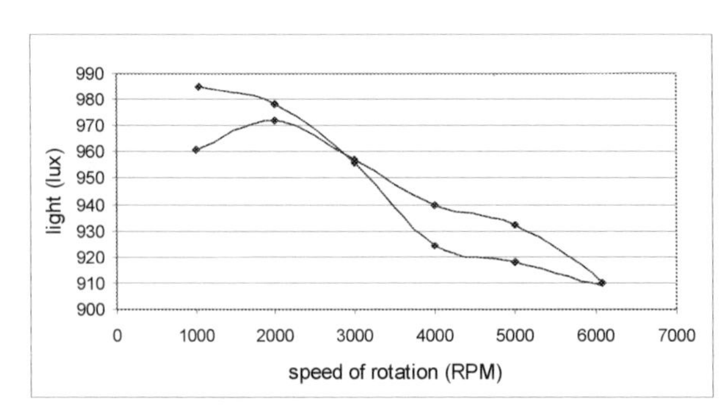

The diagram shows the strength of the light impact of 20 LED-s, function of the speed of the rotation during speeding up and slowing down
(current is: 14.1-14.9 mA)

light impact difference relative to the stationary status (lux)
0
-20
-40
-60
-80
51.2
56.2
61.2
66.2
potential (V) at 6080 RPM

The difference in the light impact at 6080 RPM relative to the stationary status also depends on the potential and the value of the current through the disc.
At the range of 50-52V and 30-100 mA within the experiment, it was around -5 lux, while at 67 V and 14 A it was -60 lux.

Diagr. 31.9

Diagram 31.9

Any "other type" of the external *blue shift* "support" has impact to the LED "light-event" as well. At certain stable voltage and current values, the impact of the temperature was recorded. Temperature demonstrates an acting and intensified *blue shift* conflict. Temperature is an increased *blue shift* conflict among molecules and atoms themselves. The increase of temperature (heating) means additional *blue shift* impact to the process. If the LED-s of the rotating disc have been heated, the light impact at the *same* original electron flow became less.

Heating decreases the potential, since its intensity increasing impact resolves part of the *blue shift* conflict, origin of the light impact. This effect results in more electron flow, but less conflict – weaker light impact.

Light impact, function of the temperature

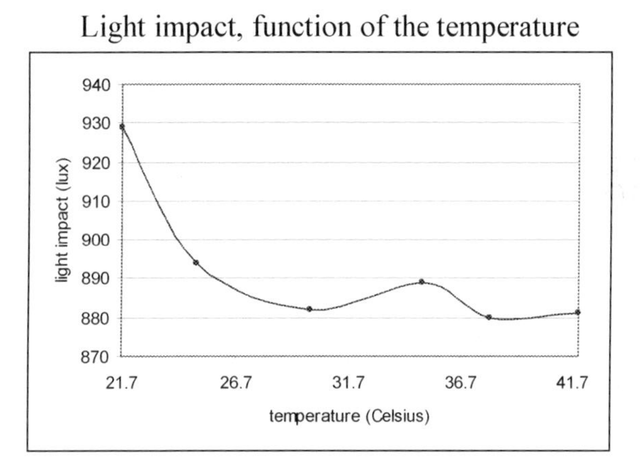

Diagram 31.10

Diagr. 31.10

The reason is that the heating provides *blue shift* impact and intensifies the neutron processes. External electron flow meets less conflict in the lighting LED devices, which means less light impact.

Higher electron flow is only increasing the light impact, if the *blue shift* conflict is the same. If part of the conflict is resolved by the heating effect, the resistance against the electron flow is less – therefore the light impact is also less. Diagram 31.10 demonstrates it.